Concepts in Physical Metallurgy

Concise lecture notes

Concepts in Physical Metallurgy

Concise lecture notes

A Lavakumar

Veer Surendra Sai University of Technology, Odisha, India

Morgan & Claypool Publishers

Rights & Permissions
To obtain permission to re-use copyrighted material from Morgan & Claypool Publishers, please contact info@morganclaypool.com.

ISBN   978-1-6817-4473-5 (ebook)
ISBN   978-1-6817-4472-8 (print)
ISBN   978-1-6817-4475-9 (mobi)

DOI   10.1088/978-1-6817-4473-5

Version: 20170401

IOP Concise Physics
ISSN 2053-2571 (online)
ISSN 2054-7307 (print)

A Morgan & Claypool publication as part of IOP Concise Physics
Published by Morgan & Claypool Publishers, 40 Oak Drive, San Rafael, CA, 94903 USA

IOP Publishing, Temple Circus, Temple Way, Bristol BS1 6HG, UK

Dedicated to my friends Ranga Nikhil Vinayak and Aksharitha Sribasyam

Contents

Preface

Developments in the science of metals have made the progress of the human race faster through the ages. In this context, the understanding of the structure–property relationship of metals and alloys has become increasingly important. The large domain of the use of materials has made material science a multidisciplinary study. This book delivers an effective way to understand and realize the behavior of the materials right from the atomic level.

The first chapter describes the need for material science in the modern age and also explains some elementary facts regarding this book. The way atoms are arranged in metallic materials and the importance of symmetry in materials are described in chapter 2. The transition of metals from liquid state to solids involves nucleation of crystals and their growth. This has been briefly covered in chapter 3. Nothing is perfect in the world as far as humans are concerned. The same applies to materials. Some possible defects exist in materials, and by altering them, one can enhance the mechanical properties of the metals. This has been given a brief overview of in chapters 4 and 5.

'United we stand, divided we fall'—this proverb denotes the strength of a collaboration. In the same manner metals require combinations in order to improve their strength and properties by the method of alloying. Chapter 6 is intended to explain the fundamentals of alloying.

Phase diagrams are the indispensable tools in the field of metallurgy. They provide a medium for understanding the different possible phases of pure metals and alloys under the influence of external agents such as temperature and pressure. Chapter 7 in this book explains the different types of phase diagrams in binary alloy systems.

Chapter 8 is about the metallurgy of ferrous metals. Ferrous metals, i.e., steels and cast irons are the most widely used alloys, and therefore it is important to understand the basics about the heat treatment techniques of steels and cast irons.

Apart from the ferrous metals, various nonferrous metals are also being used for various applications. The last chapter, chapter 9, gives a brief idea about some commonly used nonferrous metals.

Acknowledgments

Throughout the process of writing this book, many individuals have taken time out to help the author. The author would like to give special thanks to his mentors Professor Viplava Kumar, Dr N Eswar Prasad, Professor Anandh Subramanian and Professor B S Murthy who have shaped the way I think as an engineer/teacher.

First, the author thanks D Narsimhachary, Srivastava, Trinath for actively participating in the feedback and editing of this book for quality improvement and the author wishes to acknowledge Soumya Sourav for sketching some of the figures for this book.

The author would extend his gratitude to Vice-Chancellor Professor E Sai Baba Reddy VSSUT, Odisha, Dr S K Badjena (Head, MME), Dr C Sasi Kumar (MANIT, Bhopal) for their constant support. Special thanks go to friends P K Singh, Nikhil Kulkarni, Abhinay, P K Katiyar, Rita for their support and encouragement during the course of this book.

The author would like to thank his colleagues and students in VSSUT, Odisha and MANIT, Bhopal for their useful discussions.

Finally, he gratefully thanks his beloved family for all their encouragement and support during the writing of this book.

A Lavakumar

Author biography

Avala Lavakumar

Avala Lavakumar is a physical metallurgist from the Department of Metallurgy and Materials Engineering, Veer Surendra Sai University of Technology, Odisha, India. He has a Bachelor's Degree in Metallurgy and Materials Engineering from the J.N.T.U (M.G.I.T) Hyderabad, India and a Master's degree in industrial metallurgy from National Institute of Technology, Durgapur. He has research experience in RCMA (Materials), D.R.D.O. (Hyderabad) and has worked as an Assistant Professor in M.A. (National Institute of Technology, Bhopal, India). He is currently working as an Assistant Professor in Veer Surendra Sai University of Technology (V.S.S.U.T), Odisha, India. His primary research areas are indentation creep, phase transformations of steels, Al-alloys and super alloys, and their mechanical behavior.

Chapter 1

Introduction

1.1 The impact of materials on progress

Humans have evolved through many metal ages. The progress that human civilizations have made can, in part, be attributed to their learning to use various kinds of metals. Thus metallurgy can be considered as an ancient form of technology that has made our way of life more sophisticated and developed.

The beginning of the Industrial Revolution caused rapid growth in the production of iron and steel. Vast infrastructural developments took place due to improvements in metallurgy and materials science. It was thus at this time that metals and alloys, in particular steel, replaced wood as the principal structural material. The rapid developments in manufacturing, and the automotive and textile industries were also the result of the new large-scale steel production.

On the eve of the twentieth century, we had copper, bronze, iron, steel, aluminum and rubber, in addition to wood, to use as structural materials. This paved the way for a series of inventions leading to a paradigm shift in road transport, from horse-drawn carriages to motorized vehicles. In the present day, innovations in materials technology have resulted in the development of light-weight alloys and composites. Automotive engine components have traditionally been made from ferrous alloys, but the emphasis on weight reduction for higher fuel efficiency has increased the use of aluminum for cylinder blocks, cylinder heads and other engine components. Some engine covers and intake manifolds are made of magnesium. Titanium is also used in the connecting rods of high-speed engines to reduce the reciprocating mass.

The aviation and aerospace industry also owes much to developments in materials science. From the first Wright Flyer to the modern jets used today, the aerospace industry has made great progress in making human transportation much easier and faster. The use of light-weight and high-strength materials has made this possible. The aerospace industry today uses various light composite materials, such as carbon fiber composites, which are used in the bodies of various aircrafts.

doi:10.1088/978-1-6817-4473-5ch1

In the present era, electronic components have become an inseparable part of life. With advances in materials science, the sizes of computers and other instruments that use of electronic circuits have been reduced significantly. The development of smaller chips has brought the whole world to our palms.

To illustrate the importance of materials science for the electronics industry, the example of the construction of various parts of a mobile phone can be given:

- *Display*. This relies upon the combination of a liquid crystal display and a touch screen for communication with the device. The touch screen is made from a conductive but transparent material, indium tin oxide, a ceramic conductor.
- *Integrated circuits* (ICs). At the heart of the iPhone are a number of ICs, built upon billions of individual transistors, all of which rely on precise control of the semiconductor material, silicon, to which dopant atoms have been added to change the silicon's electronic properties
- *Interconnects*. The interconnects, which provide the links between components, are now made of copper, not aluminum, for higher speed and efficiency.
- *Wireless*. Microwave circuits need capacitors, which are ceramic insulators whose structure and composition is carefully controlled to optimize the capacitance.
- *Battery*. The battery is a modern Li-ion battery where the atomic structure of the electrodes is carefully controlled to enable the diffusion of the Li ions.
- *Headphones*. Most headphones use modern magnetic materials, whose structure and composition has been developed to produce very strong permanent magnets. This forms part of a transducer that turns electrical signals into sound.

1.2 A possible classification of physical metallurgy

Figure 1.1 shows a broad classification of physical metallurgy. The 'physical' category will be the topic of this book, explaining the structure property correlation of the different classes of materials.

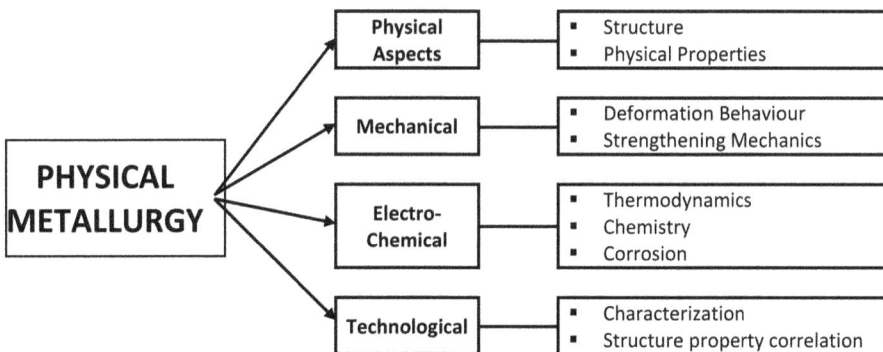

Figure 1.1. A broad classification of physical metallurgy.

1.3 Electrons to components

In the early scientific literature, the atom was considered to be the smallest, chemically indivisible particle of matter. The smallest quantity of a substance which can exist freely by itself in a chemically recognizable form was known as a molecule. It was only in 1803 that John Dalton published his famous theory, the concept of atomicity. In this theory, he suggested that molecules are composed of atoms of different elements in a fixed proportion. In 1897, J J Thomson, while studying the passage of electricity through the low-pressure gases, discovered that gases ionize into positive and negative charges. Several studies, then showed the presence of various smaller particles known as fundamental particles.

The fundamental particles inside the atoms that are important from the point of view of our subject are:

- *Electrons*. These are negatively charged particles. The mass of an electron is 9.1×10^{-31} kg, which is equal to 1/1836 of the mass of a hydrogen atom. Each possesses a unit negative charge of electricity, which is equal to 1.602×10^{-19} coulombs (C).
- *Protons*. These are positively charged particles. The mass of a proton is 1.672×10^{-27} kg. Each proton possesses a unit positive charge of electricity; this charge is also equal to 1.602×10^{-19} C.
- *Neutrons*. These are electrically neutral particles. The mass of a neutron is 1.675×10^{-27} kg. This is approximately equal to the mass of a proton. Each neutron is composed of one proton and one electron.

As Richard Feynman said, 'It would be very easy to make an analysis of any complicated chemical substance; all one would have to do would be to look at it and see where the atoms are …'. And by modifying those positions in different length scales, we can simply improve the properties of the materials. The possible length scale in this field is given below:

1. Electronic structure
2. Atomic structure
3. Crystal structure

Table 1.1. Different classes of the properties of materials.

Economic	Price and availability, recyclability
General physical	Density
Mechanical	Modulus, yield and tensile strength, hardness, fracture strength, fatigue strength, creep strength, damping
Thermal	Thermal conductivity, specific heat
Electric and magnetic	Resistivity, dielectric constant, magnetic permeability
Environmental interactions	Oxidation, corrosion and wear
Production	Ease of manufacture, joining, finishing
Aesthetic (appearance)	Color, texture, feel

4. Microstructure
5. Macrostructure
6. Component

This leads to selecting a suitable material for a component by considering the various factors and properties (table 1.1).

Further reading

Askeland D R and Phulé P 2006 *The Science and Engineering of Materials* (Boston, MA: Cengage Learning)

Avner S H 1997 *Introduction to Physical Metallurgy* (India: McGraw-Hill)

Callister W D 2007 *Callister's Materials Science and Engineering* (Indian Adaptation, adapted by R Balasubramaniam) (New Delhi: Wiley)

Raghavan V 2004 *Materials Science and Engineering* 5th edn (Englewood Cliffs, NJ: Prentice-Hall)

Subramaniam A and Balani K (IITK) *Materials Science and Engineering* (e-book) MHRD, India

Chapter 2

Crystal structures

Solids are either crystalline or non-crystalline. The majority of engineering materials, many ceramics, most minerals, some plastics and all metals are crystalline in structure. The type of crystal structure has a significant bearing on the physical properties of these materials. The various defects which arise in the formation of the crystals of a material are further responsible for certain aspects of the material's chemical and physical behavior.

2.1 Platonic solids

In 387 BCE a philosopher called Plato, the most famous student of Socrates, believed that the entire Universe is made up of five elements: fire, earth, air, water, and god(s). In his view the whole Universe was a periodic arrangement of five solids, now called the Platonic solids (see figure 2.1). According to his argument, in the Universe the Platonic solids represent the elements as follows:

- tetrahedron (4 faces)—fire
- hexahedron (cube) (6 faces)—earth
- octahedron (8 faces)—air
- dodecahedron (12 faces)—god
- icosahedron (20 faces)—water

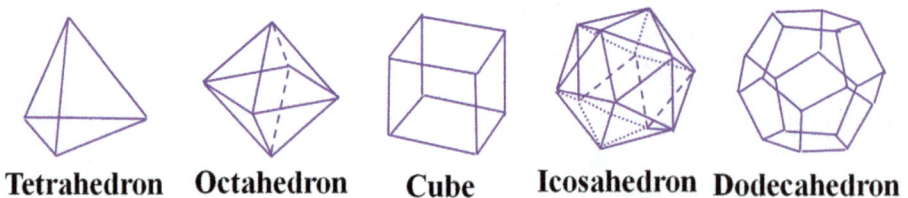

Tetrahedron   Octahedron   Cube   Icosahedron Dodecahedron

Figure 2.1. Different platonic solids.

Icosahedrons will provide the best packing, but if we translate this into three dimensions we cannot obtain a continuous structure without any voids. This presence of voids in crystallography is called frustration. Previously, people thought it was not possible to have a solid with icosahedron packing; however, it proved to be possible when quasicrystals (Al–Mn) were discovered by Shechtman in 1984.

2.2 The crystal, lattice and motif

A three-dimensional translationally periodic arrangement of atoms in space is called a crystal, and a three-dimensional translationally periodic arrangement of points in space is called a lattice. The terms have a similar meaning but describe different entities. A crystal is finite but a lattice might be infinite. The interconnection between these two terms can be explained by the concept of a motif or basis, which can be defined as, typically, *an atom or a group of atoms associated with each lattice point.*

$$\text{crystal} = \text{lattice} + \text{motif}.$$

So the lattice determines *how* to repeat and the motif determines *what* to repeat.

The set of these lattice points constitutes a three-dimensional lattice. A unit cell may be defined within this lattice as a space-filling parallelepiped with its origin at a lattice point, and with its edges given by three non-coplanar basis vectors a_1, a_2 and a_3, each of which represents translations between two adjacent lattice points. The entire lattice can then be generated by stacking unit cells in three dimensions. Any vector representing a translation between lattice points is called a lattice vector.

The unit cell defined above has lattice points located at its corners. Since these are shared with seven other such cells, and since each cell has eight corners, there is only one lattice point per unit cell. Such a unit cell is primitive and has the lattice symbol P.

Non-primitive unit cells can have two or more lattice points, in which case the additional ones will be located at positions other than the corners of the cell.

The basis vectors x, y and z define the unit cell. Their magnitudes a, b and c, respectively, are the lattice parameters of the unit cell. The angles x–y, y–z and z–x are conventionally labeled, α, β and γ, respectively (figure 2.2).

2.3 The concept of symmetry

Symmetry can be defined as follows: if an object is brought into self-coincidence after some operation, it is said to possess a symmetry with respect to that operation. Symmetry in this context refers to geometrical symmetry. As symmetry manifests the inherent, external shape of a crystal, this concept helps to reduce the infinite amount of information needed to describe a crystal into a finite amount of information. In fact, crystals are defined on the basis of symmetry only. The symmetry under consideration could be that of one of the following entities: the lattice, crystal, motif and unit cell.

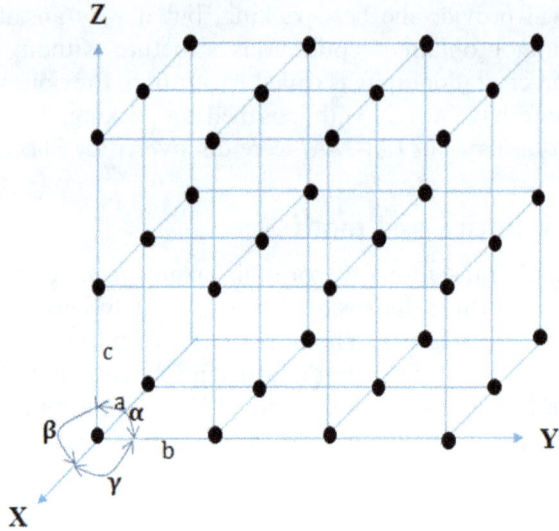

Figure 2.2. The lattice parameters and angles between the faces.

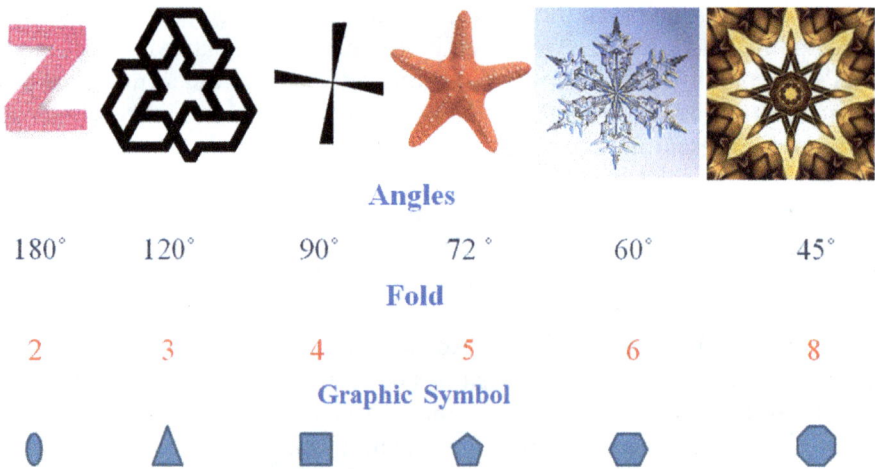

Angles

| 180° | 120° | 90° | 72° | 60° | 45° |

Fold

| 2 | 3 | 4 | 5 | 6 | 8 |

Graphic Symbol

Figure 2.3. Some symmetry operations.

Although the properties of a crystal can be anisotropic, there may be different directions along which they are identical. These directions are said to be equivalent and the crystal is said to possess symmetry. Broadly, symmetry operations are two types. These are translational and rotational. Some symmetry operations are illustrated in figure 2.3; in essence, they transform a spatial arrangement into another which is indistinguishable from the original.

An object possesses an n-fold axis of rotational symmetry if it coincides with itself upon rotation about the axis through an angle of $360°/n$. The possible angles of rotation, which are consistent with the translational symmetry of the lattice, are

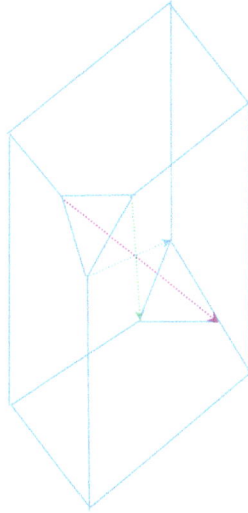

Figure 2.4. Illustration of center of symmetry and inversion. The center of symmetry is at the point where the arrows cross. Reproduced with permission of Professor Sir H Bhadeshia, Cambridge University.

360°, 180°, 120°, 90° and 60° for values of n equal to 1, 2, 3, 4 and 6, respectively. A five-fold axis of rotation does not preserve the translational symmetry of the lattice and hence is forbidden. A one-fold axis of rotation is called a monad and the terms diad, triad, tetrad and hexad correspond to $n = 2$, 3, 4 and 6, respectively. All of the Bravais lattices have a center of symmetry (figure 2.4). An observer at the center of symmetry sees no difference in the arrangement between the directions.

Symmetry operations generate a variety of arrangements of lattice points in three dimensions. There are 32 unique ways in which lattice points can be arranged in space. These non-translation elements are called *point groups.* A large number of three-dimensional structures are generated when translations (linear translation, translation + reflection (glide plane) and translation + rotation (screw axis)) are applied to the point groups. There are 230 unique shapes which can be generated this way. These are called *space groups.*

The 32 point groups are denoted by notations called Hermann–Mauguin symbols. These symbols describe the unique symmetry elements present in a body. The shape in figure 2.5 contains one four-fold axis, four two-fold axes and five mirror planes. Three of the mirror planes and two of the two-fold axes are unique, as the others can be produced by a symmetry operation. Therefore, the point group symbol in this shape is 4/m2/m2/m. The '/' between 4 or 2 and 'm' indicates that they are perpendicular to each other. The symmetry of the cube is given in figure 2.6.

2.4 Bravais lattices

The fourteen distinguishable three-dimensional space lattices that can be generated by repeated translation of three non-coplanar vectors, a, b and c, of a unit cell in

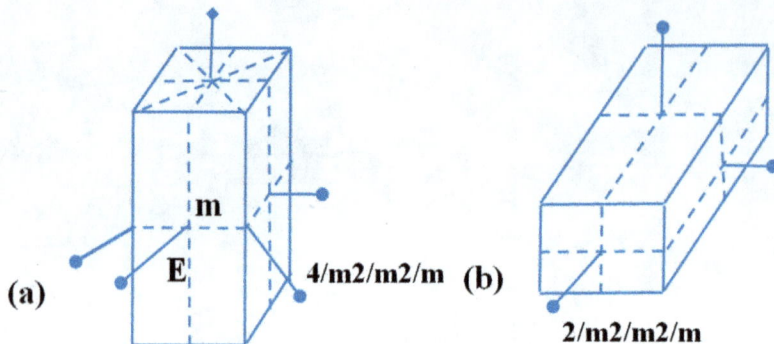

Figure 2.5. Hermann–Mauguin symbols for different symmetry operations.

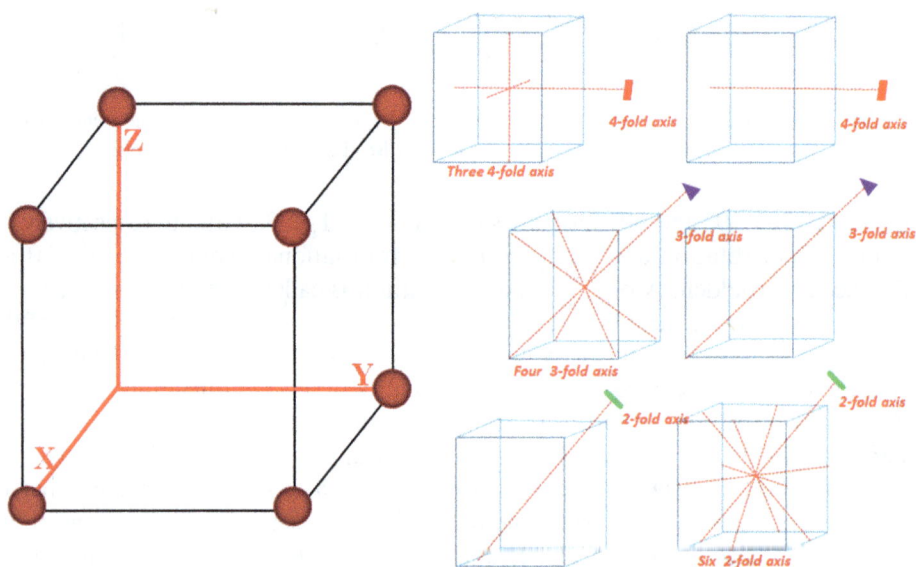

Figure 2.6. Symmetry of a cube.

three-dimensional space are known as Bravais lattices, named after their originator. Figure 2.7 shows conventional unit cells of 14 Bravais lattices. Table 2.1 lists the lattice parameters of the crystal systems.

2.5 Miller indices for planes and directions

2.5.1 Miller indices for planes

Miller indices are used to specify directions and planes. These directions and planes could be in lattices or in crystals. The number of indices will match the dimensions of the lattice or, in crystal: in one dimension there will be one index and in two dimensions there will be two indices, etc. Some aspects of Miller indices, in particular those for planes, are not intuitively understood.

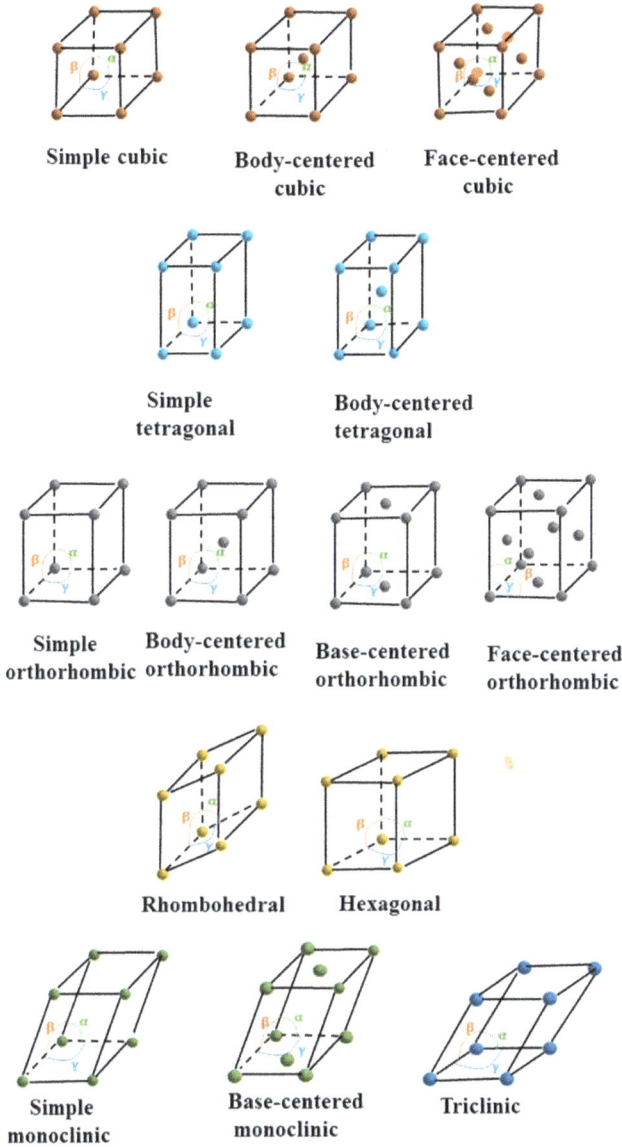

Figure 2.7. Different crystal structures.

The Miller indices of a plane, indicated by h, k and l, are given by the reciprocal of the intercepts of the plane on the three axes. A plane which intersects the x-axis at 1 (one lattice parameter) and is parallel to the y- and z-axes, has the Miller indices $h = 1/1 = 1$, $k = 1/\infty = 0$, $l = 1/\infty = 0$. This is written as $(hkl) = (100)$.

To find the Miller indices of a plane, follow these steps:
- Determine the intercepts of the plane along the crystal axes.
- Take the reciprocals.

Table 2.1. Crystal systems and lattice parameters. P: primitive; I: body-centered; F: face-centered; C: end-centered.

Crystal system	Lattice parameters	Bravais lattices			
		P	I	F	C
1 Cubic	$(a = b = c, \alpha = \beta = \gamma = 90°)$	✓	✓	✓	
2 Tetragonal	$(a = b \neq c, \alpha = \beta = \gamma = 90°)$	✓	✓		
3 Orthorhombic	$(a \neq b \neq c, \alpha = \beta = \gamma = 90°)$	✓	✓	✓	✓
4 Hexagonal	$(a = b \neq c, \alpha = \beta = 90°, \gamma = 120°)$	✓			
5 Trigonal	$(a = b = c, \alpha = \beta = \gamma \neq 90°)$	✓			
6 Monoclinic	$(a \neq b \neq c, \alpha = \gamma = 90° \neq \beta)$	✓			✓
7 Triclinic	$(a \neq b \neq c, \alpha \neq \beta \neq \gamma)$	✓			

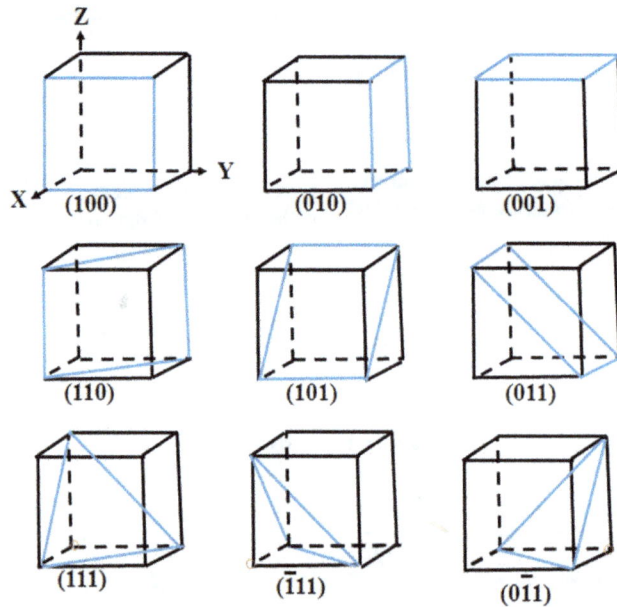

Figure 2.8. Illustration of different planes in a cubic crystal.

- Clear fractions.
- Reduce to the lowest terms, and enclose in brackets ( ).

The Miller indices of the planes in a cubic crystal are shown in figure 2.8.

2.5.2 Families of planes

A set of planes related by symmetry operations of the lattice or the crystal is called a family of planes. Planes with similar indices are equivalent, e.g. the faces of the cube (100), (010) and (001). This is termed as a family of planes and is denoted as {100}, which includes all the (100) combinations including negative indices.

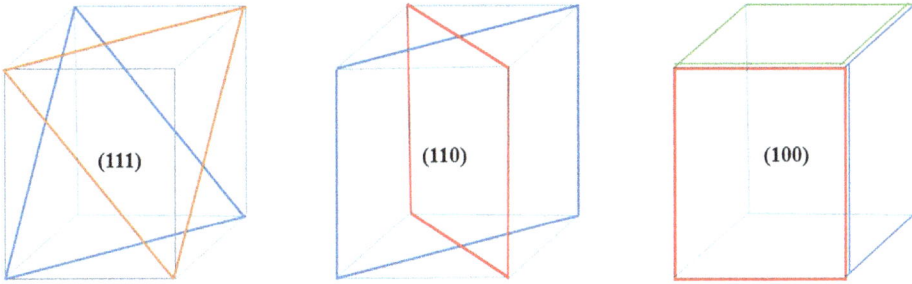

Figure 2.9. Family of planes in a cubic crystal.

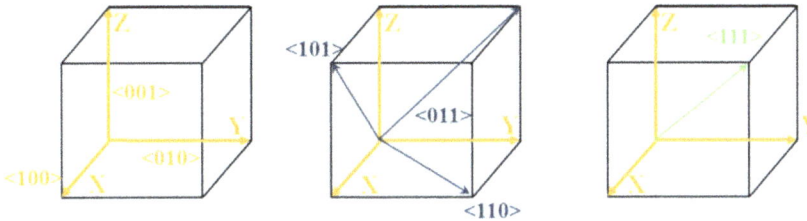

Figure 2.10. Directions in a cubic crystal system.

All the points which one should keep in mind while dealing with directions to obtain the members of a family, should also be kept in mind when dealing with planes. The family of planes for a cubic crystal is given in figure 2.9.

2.5.3 Miller indices for crystal directions

The directions in a crystal are given by specifying the coordinates (u, v, w) of a point on a vector (r_{uvw}) passing through the origin. $r_{uvw} = ua + vb + wc$, which is indicated as $[uvw]$. For example, the direction $[110]$ lies on a vector r_{110}, whose projection lengths on the x- and y-axes are one unit (in terms of unit vectors a and b).

To determine a direction of a line in the crystal:

- Find the coordinates of the two ends of the line and subtract the coordinates (head–tail) *or* draw a line from the origin parallel to the line and find its projection lengths on the x-, y- and z-axes in terms of the unit vectors a, b and c.
- Convert fractions, if any, into integers and reduce to the lowest term.
- Enclose in square brackets $[uvw]$.

Different directions drawn in a cubic crystal are shown in figure 2.10.

2.5.4 Families of directions

A set of directions related to symmetry operations of the lattice or the crystal is called a family of directions. A family of directions is represented (in Miller index notation) as $\langle u\ v\ w \rangle$ (see figure 2.11).

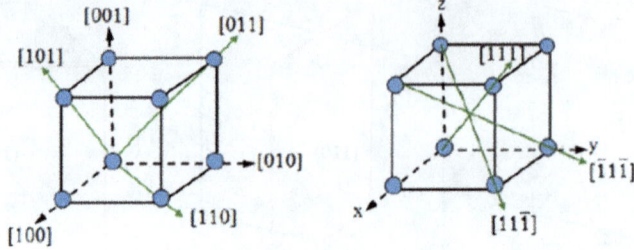

Figure 2.11. Family of directions in a cubic crystal system.

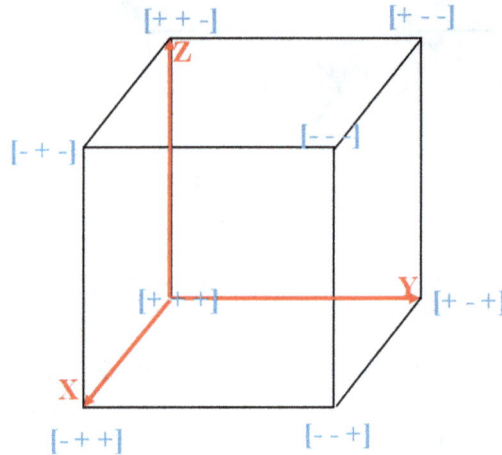

Figure 2.12. Unit cell with points denoted by positive and negative indices.

A Lava's way to understand negative indices

Planes and directions can also have negative indices. To identify negative indices, we should introduce another unit cell on the negative side of the axis, but this may lead to complications. Therefore, drawing in a single unit cell is made much easier by shifting the center of the indices to a suitable one (see figure 2.12):

1. Draw a unit cell and identify the indices using the above-described method, and then fix the origin for the given indices.
2. Follow the above-described rules of planes and directions and determine the Miller indices, when there are any negative indices.

2.5.5 The hexagonal system

Planes

In the cubic system all the faces of the cube are equivalent, that is, they have similar indices. However, this is not the case in the hexagonal system. The six prism faces, for example, have indices (100), (010), ($\bar{1}$10), ($\bar{1}$00), (0$\bar{1}$0), (1$\bar{1}$0), which are not the same. In order to address this, a fourth axis (a_3), which is opposite to the vector sum

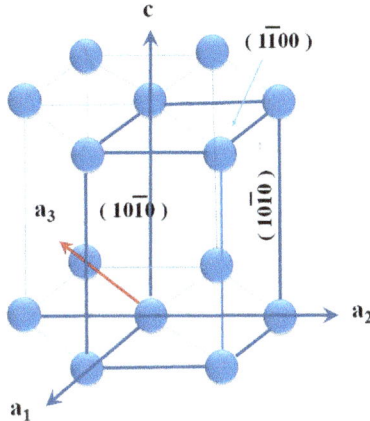

Figure 2.13. Miller indices of the faces in a hexagonal system.

of a_1 and a_2, is used and a corresponding fourth index i is used along with hkl. Therefore, the indices of a plane are given by $(hkil)$ where $i = -(h + k)$. Sometimes i is replaced by a dot and written as $(hk.l)$ (see figure 2.13).

The indices of six faces now become $(10\bar{1}0)$, $(01\bar{1}0)$, $(\bar{1}100)$ and $(0\bar{1}10)$, which are now equivalent and belong to the $\{10\bar{1}0\}$ family of planes.

Directions

Like planes, directions in the hexagonal system are also written in terms of four indices as $[uvtw]$. If $[UVW]$ are the indices in three axes, this can be converted to four-axis indices $[uvtw]$ using the following relations:

$$U = u - t \qquad V = v - t \qquad W = w$$

$$u = \frac{2U - V}{3} \qquad v = \frac{2V - U}{3} \qquad t = -(u + v) = -\frac{U + V}{3} \qquad w = W$$

Ex: $[100] = [2\bar{1}\bar{1}0]$, $[210] = [10\bar{1}0]$.

2.6 The coordination number

The coordination number (CN), Z, is defined as the total number of nearest neighbor atoms:

- *Simple cubic* (sc). The CN of an sc crystal is 6. This is shown in figure 2.14, in which we consider one corner and draw three lines connecting six points.
- *Body-centered cubic* (bcc). The CN of a bcc crystal is 8. The body-centered atom is in contact with all eight corner atoms. Each corner atom is shared by eight unit cells and, hence, each of these atoms is in contact with eight body-centered atoms.
- *Face-centered cubic* (fcc). In the fcc lattice each atom is in contact with 12 neighboring atoms. The fcc CN is $Z = 12$. For example, the face-centered atom in the front face is in contact with four corner atoms and four other

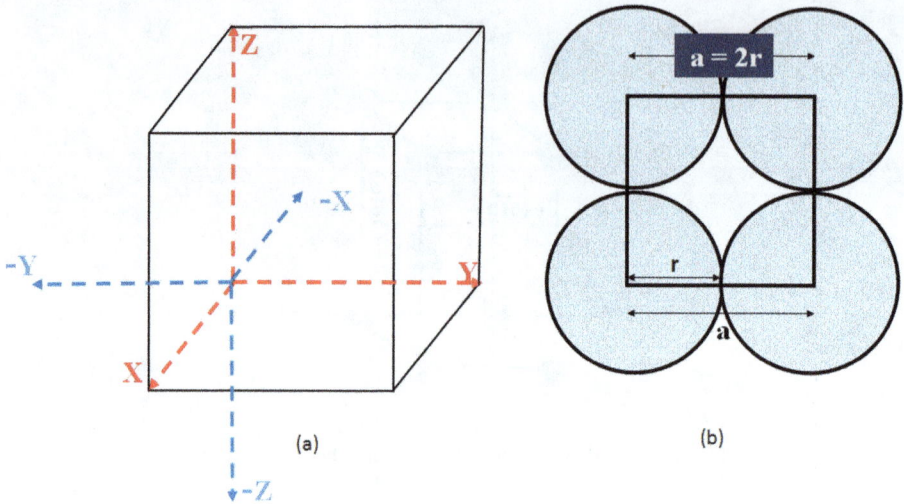

Figure 2.14. Simple cubic unitcell showing (a) coordination number and (b) packing fraction.

Table 2.2. The average number of atoms per unit cell.

		Position of atoms	Effective number of atoms
1	sc	8 corners	$= [8 \times (1/8)] = 1$
2	bcc	8 corners + 1 body centre	$= [1 \text{ (for corners)}] + [1 \text{ (BC)}] = 2$
3	fcc	8 corners + 6 face centre	$= [1 \text{ (for corners)}] + [6 \times (1/2)] = 4$
4	hcp	12 corners (6 bottom + 6 top) + 2 atoms at face centers + 3 atoms in the interior	$= [12 \times (1/6)] + [2 \times (1/2)]$ $+ [3 \text{ (interior)}] = 6$

face-centered atoms behind it (two sides, top and bottom) and is also touching four face-centered atoms of the unit cell in front of it.

- *Hexagonal close-packed* (hcp). In a hexagonal lattice $Z = 12$. The center atom of the top face is in contact with six corner atoms, three atoms of the mid-layer and three atoms of the mid-layer of the unit cell above it. Table 2.2, summarizes the average number of atoms per unit cell in sc, bcc, fcc and hcp.

2.7 The atomic packing factor

The atomic packing factor (APF) or packing efficiency indicates how closely atoms are packed in a unit cell and is given by the ratio of the volume of atoms in the unit cell and the volume of the unit cell:

$$\text{APF} = \frac{\text{volume of atoms}}{\text{volume of unit cell}}.$$

2.7.1 Simple cubic

In an SC structure, the atoms are assumed to be placed in such a way that any two adjacent atoms touch each other. If a is the lattice parameter of the sc structure and R the radius of the atoms, from figure 2.14 it is clear that atomic radius $(R) = a/2$

$$\text{APF} = \frac{\text{average number of atoms per unit cell} \times \text{volume of an atom}}{\text{volume of the unit cell}}$$

$$\text{APF}_{sc} = \frac{1 \times \frac{4}{3} \Pi R^3}{a^3} = \frac{\frac{4}{3} \Pi R^3}{(2R)^3} = \frac{4\Pi}{24} \times 100 = 0.52 = 52\%.$$

2.7.2 Body-centered cubic

In bcc structures the center atom touches the corner atoms as shown in figure 2.15:

$$c^2 = b^2 + a^2 \rightarrow c^2 = 2a^2 + a^2$$
$$b^2 = a^2 + a^2 = 2a^2$$

$$c^2 = 3a^2 \rightarrow (4R)^2 = 3a^2 \rightarrow R^2 = \frac{3a^2}{16}$$

$$R = \frac{\sqrt{3}a}{4}, \ a = \frac{4R}{\sqrt{3}}$$

$$\text{APF} = \frac{2 \times \frac{4}{3} \Pi R^3}{a^3} = \frac{2 \times \frac{4}{3} \Pi \left(\frac{\sqrt{3}a}{4}\right)^3}{a^3} = \frac{8 \times 3\sqrt{3} \Pi a^3}{3 \times 64a^3}$$

$\text{APF} = 0.68$ (or 68%).

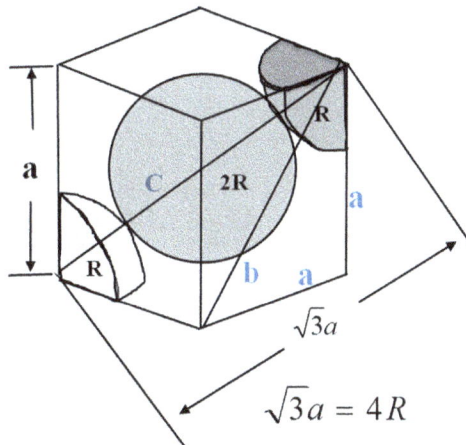

Figure 2.15. A bcc unit cell with lattice parameter a and radius R.

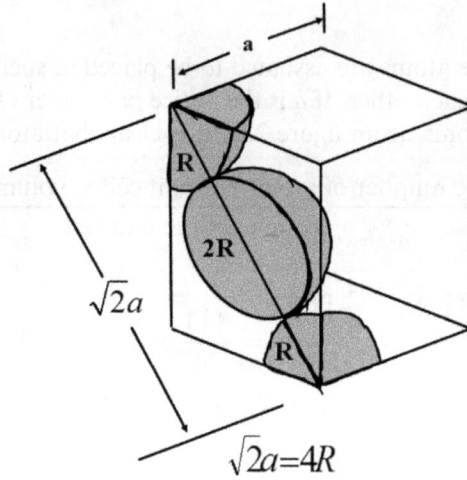

Figure 2.16. FCC unit cell with lattice parameter '*a*' and atomic radius *R*.

2.7.3 Face-centered cubic

In fcc structures the face atom touches the corner atoms as shown in figure 2.16. From this figure we can first calculate the atomic radius:

$$(4R)^2 = a^2 + a^2 \rightarrow 16R^2 = 2a^2 \rightarrow R^2 = \frac{2a^2}{16}$$

$$R = \frac{a\sqrt{2}}{4} = \frac{a}{2\sqrt{2}}$$

$$a = \frac{4r}{\sqrt{2}} \rightarrow 4R = \sqrt{2}a.$$

Then

$$\text{APF} = \frac{4 \times \frac{4}{3}\Pi R^3}{a^3} = \frac{4 \times \frac{4}{3}\Pi \left(\frac{a}{2\sqrt{2}}\right)^3}{a^3} = \frac{16 \times 2\sqrt{2}\,\Pi a^3}{3 \times 64a^3} = 0.74$$

$$\text{APF}_{\text{fcc}} = 0.74 \text{ (or 74\%)}.$$

2.7.4 Hexagonal close-packed structure

Consider any one triangle. Let us consider $\triangle AOB$; P is the center of triangle $APOB$, a tetrahedron (figure 2.17).

In $\triangle AYB$

$$\cos 30° = (AY)/(AB).$$

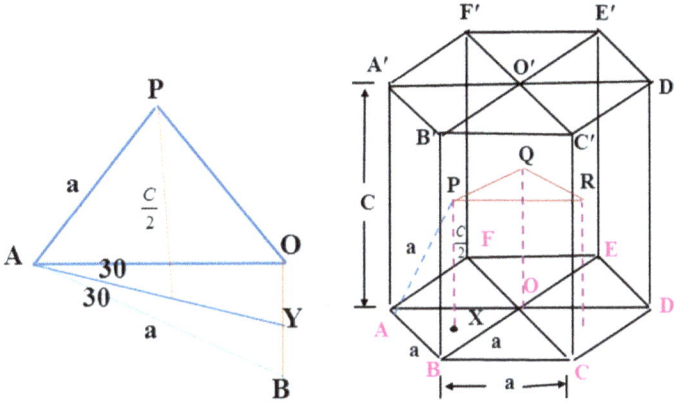

Figure 2.17. An HCP unit cell.

The distance between any neighboring atoms is a from $AB = a$:

$$AY = AB \cos 30° = \frac{a\sqrt{3}}{2}.$$

From figure 2.17, Ax is the orthocenter, so

$$Ax = \frac{2}{3}AY = \frac{2}{3} \times \frac{a\sqrt{3}}{2} = \frac{a}{\sqrt{3}}$$

$$(AP)^2 = (Ax)^2 + (xP)^2$$

$$a^2 = \frac{a^2}{3} + \frac{c^2}{4} \rightarrow a^2 - \frac{a^2}{3} = \frac{c^2}{4} \rightarrow \frac{c^2}{4} = \frac{2a^2}{3}$$

$$\mathrm{APF} = \frac{\text{average number of atoms per unit cell} \times \text{volume of an atom}}{\text{volume of the unit cell}}$$

$$\mathrm{APF} = \frac{6 \times \frac{4}{3}\Pi R^3}{\text{volume of the unit cell}}$$

In cubic structures $a = b = c$ so the volume is a^3, but in the case of hcp $a = b \neq c$:

volume of unit cell = area of base of hexagonal × height

volume of unit cell = $6 \times$ area of $\triangle AOB \times c$

volume of unit cell = $6 \times \dfrac{1}{2} \times OB \times AY \times c$

volume of unit cell = $6 \times \dfrac{1}{2} \times a \times \dfrac{a\sqrt{3}}{2} \times c = ca^2\dfrac{3\sqrt{3}}{2}$

$$APF = \frac{6 \times \frac{4}{3}\Pi R^3}{ca^2 \frac{3\sqrt{3}}{2}}$$

$$APF = \frac{6 \times \frac{4}{3}\Pi R^3}{ca^2 \frac{3\sqrt{3}}{2}} = \frac{6 \times \frac{4}{3}\Pi \times \frac{a^3}{8}}{ca^2 \frac{3\sqrt{3}}{2}} = \frac{2\Pi a}{c \times 3\sqrt{3}} = 0.74$$

$APF_{hcp} = 0.74$ (or 74%).

2.7.5 Diamond cubic (dc) structure

Consider the triangle WXY from figure 2.18

$$XY^2 = XW^2 + WY^2$$
$$= \left(\frac{a}{4}\right)^2 + \left(\frac{a}{4}\right)^2$$
$$XZ^2 = \frac{a^2}{8}$$

Also in the triangle XYZ,

$$XZ^2 = XY^2 + YZ^2$$
$$= \frac{a^2}{8} + \left(\frac{a}{4}\right)^2$$
$$XZ^2 = \frac{3a^2}{16}$$

But $XZ = 2r$,

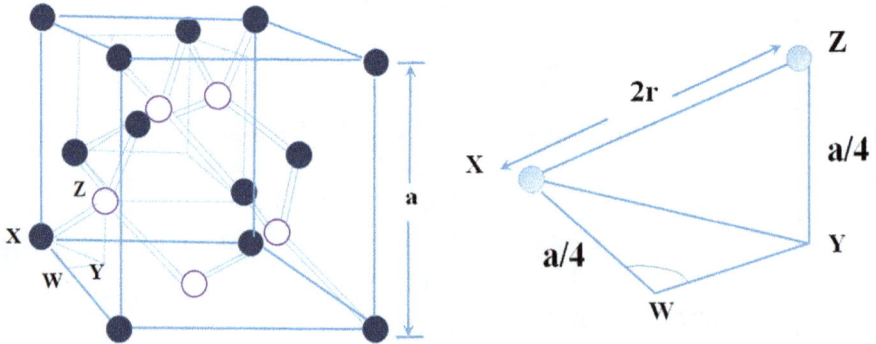

Figure 2.18. A dc unit cell.

Table 2.3. Summary of packing fraction/efficiency.

	sc	bcc	fcc	dc	hcp
Relation between atomic radius R and lattice parameter a	$a = 2R$	$\sqrt{3}\,a = 4R$	$\sqrt{2}\,a = 4R$	$\dfrac{\sqrt{3}}{4}a = 2R$	$a = 2$ $c = 4R\sqrt{\dfrac{2}{3}}$
Atoms/cell	1	2	4	8	2
Lattice points/cell	1	2	4	4	1
No. of nearest neighbours	6	8	12	4	12
Packing fraction	$\dfrac{\pi}{6}$	$\dfrac{\sqrt{3}\pi}{8}$	$\dfrac{\sqrt{2}\pi}{6}$	$\dfrac{\sqrt{3}\pi}{16}$	$\dfrac{\sqrt{2}\pi}{6}$
	~ 0.52	~ 0.68	~ 0.74	~ 0.34	~ 0.74

$$\therefore (2r)^2 = \frac{3a^2}{16} \Rightarrow r^2 = \frac{3a^2}{64} \Rightarrow r = \frac{\sqrt{3}}{8}a$$

$$\text{APF} = \frac{\text{volume of atoms}}{\text{volume of unitcell}}$$

$$= \frac{8 \times \frac{4}{3}\Pi r^3}{a^3}$$

$$= \frac{8 \times \frac{4}{3}\Pi \left(\frac{\sqrt{3}}{8}a\right)^3}{a^3}$$

$$= \frac{8 \times 4\Pi \times 3\sqrt{3}\,a^3}{3 \times a^3 \times 8^3}$$

$$\text{APF} = 0.34 \text{ or } 34\%.$$

Table 2.3 summarizes the atomic packing factor of sc, bcc, fcc, dc and hcp structures.

2.8 Density calculations

2.8.1 Linear density

Linear density (LD) is the number of atoms per unit length along a particular direction

$$\text{LD} = \frac{\text{number of atoms on the direction vector}}{\text{length of the direction vector}}.$$

From figure 2.19, the $\langle 110 \rangle$ directions in the fcc lattice have two atoms (1/2 × 2 corner atoms + 1 center atom) and the length is

$$\text{LD}_{[110]} = \frac{2}{\sqrt{2}a} = \frac{\sqrt{2}}{a}.$$

This is the most densely packed direction in the fcc lattice.

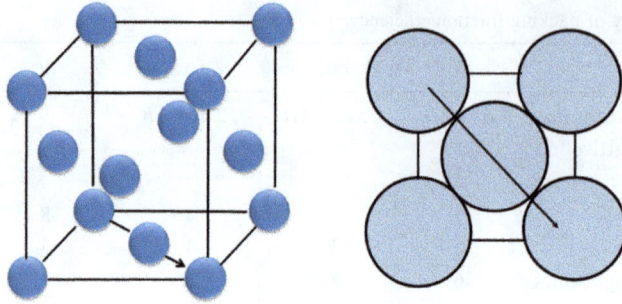

Figure 2.19. An fcc unit cell.

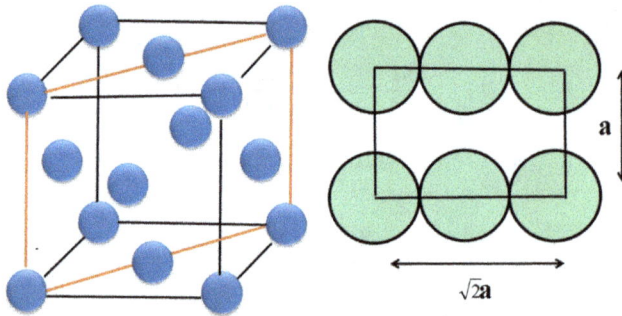

Figure 2.20. {110} planes in an fcc crystal.

2.8.2 Planar (areal) density

Planar density (PD) refers to the density of atomic packing on a particular plane:

$$\text{PD} = \frac{\text{number of atoms on a plane}}{\text{area of plane}}.$$

For example, there are two atoms $(1/4 \times 4$ corner atoms $+ 1/2 \times 2$ side atoms) in the {110} planes in the fcc lattice. The PD of {110} planes in the fcc crystal is (figure 2.20):

$$\text{PD}_{(110)} = \frac{2}{a\sqrt{2a}} = \frac{\sqrt{2}}{a^2}.$$

In the {111} planes of the fcc lattice there are two atoms $(1/6 \times 3$ corner atoms $+ 1/2 \times 3$ side atoms). The PD of the {111} planes in the fcc crystal are

$$\text{PD}_{(111)} = \frac{2}{\frac{1}{2}\sqrt{2}a \times \sqrt{2}a\frac{\sqrt{3}}{2}} = \frac{4}{\sqrt{3}a^2}.$$

This is higher than {110} and any other plane. Therefore, {111} planes are the most densely packed planes in the fcc crystal (figure 2.21).

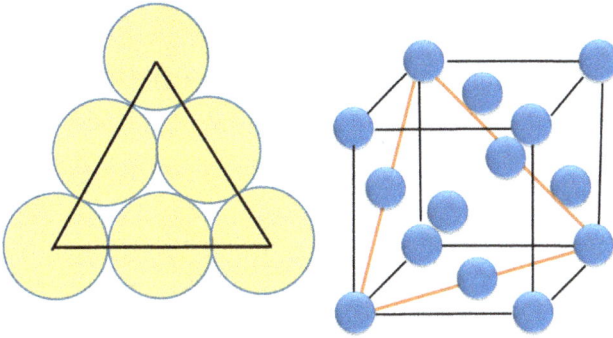

Figure 2.21. The {111} plane in an fcc crystal.

2.8.3 Theoretical (volume) density

The equation for calculating the theoretical density, ρ, from crystal structure is as follows:

$$\rho = \frac{nA}{V_C N_A}$$

where n is the number of atoms in the unit cell, A is atomic weight, V_C is the volume of the unit cell and N_A = Avogadro's number (6.023×10^{23}).

So how would one calculate the theoretical density of aluminum (Al)? Given that Al is an fcc structure with the lattice parameter 4.05 Å, $n = 4$ and an atomic weight of 26.98 g/mol,

$$\rho = \frac{4 \times 26.98}{(4.05 \times 10^{-8})^3 \times 6.023 \times 10^{23}} = 2.697\,\text{g/cc}.$$

2.9 Structure–property correlation

Al is ductile while iron (Fe) and magnesium (Mg) are not. This can be explained from their crystal structures. Al is fcc, whereas Fe is bcc and Mg is hcp. Plastic deformation in metals takes place mainly by a process called slip. Slip can be broadly visualized as the sliding of crystal planes over one another. Slip mostly occurs on densely packed planes in the most closely packed directions lying on that plane. The slip plane and the direction together are called a slip system.

In fcc structures, the {111} planes are close-packed and there are four unique {111} planes. Each of these planes contains three closely packed $\langle 110 \rangle$ directions. Therefore, there are $4 \times 3 = 12$ slip systems. In hcp structures, the basal plane, (0001), is close-packed and it contains three $\langle 11\underline{2}0 \rangle$ directions. Hence, the number of slip systems is $1 \times 3 = 3$. Slip in a larger number of slip systems allows greater plastic deformation before the fracture, imparting ductility to fcc materials. Close-packed planes are also the planes with the greatest interplanar spacing and this allows slip to take place easily on these planes.

The bcc structures on the other hand have 48 possible slip systems. However, there is no close-packed plane. Hence, plastic deformation before fracture is not significant. Slip might occur in the {110}, {112} and {123} planes in the ⟨111⟩ directions.

2.10 Voids in crystals

We have already seen that as spheres cannot fill the entire space, the packing fraction (PF) is smaller than 1 (for all crystals). This implies there are voids between the atoms. The lower the PF, the larger the volume occupied by voids. These voids have complicated shapes; but we are mostly interested in the largest sphere which can fit into these voids, hence only the plane-faced polyhedron version of the voids is (typically) considered. The size and distribution of voids in materials play a role in determining aspects of material behavior, e.g. the solubility of interstitials and their diffusivity.

In the close-packed crystals (fcc, hcp) there are two types of voids, tetrahedral and octahedral. These are identical in both structures as the voids are formed between two layers of atoms. In a bcc crystal the voids do *not* have the shape of a regular tetrahedron or regular octahedron. The octahedral and tetrahedral voids of FCC and BCC are given in figure 2.22.

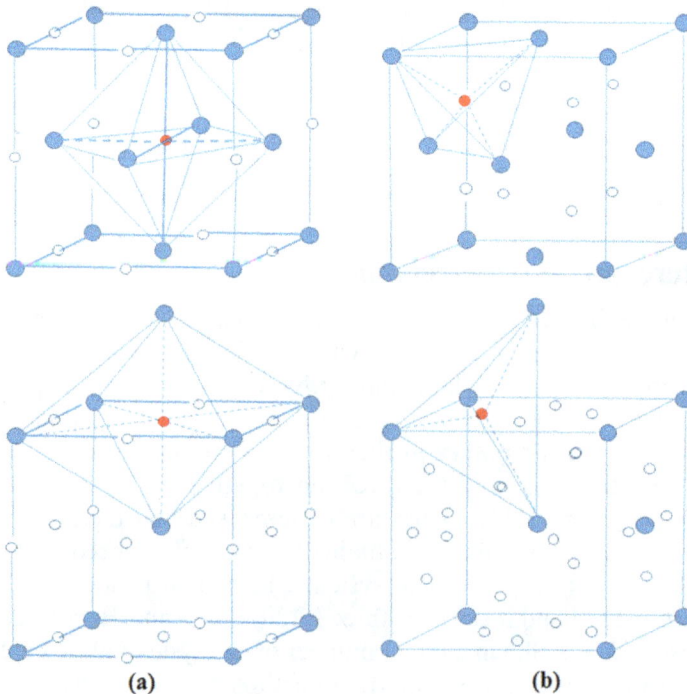

(a)   (b)

Figure 2.22. (a) Octahedral voids of FCC and BCC. (b) Tetrahedral voids of FCC and BCC.

Further reading

American Society for Testing and Materials 1971 *Annual Book of Standards* (Philadelphia, PA: ASTM)

Askeland D R and Phulé P 2006 *The Science and Engineering of Materials* (Boston, MA: Cengage Learning)

Avner S H 1997 *Introduction to Physical Metallurgy* (New York: McGraw-Hill)

Bauri R 2001 Introduction to Material Science and Engineering, IIT Madras NPTEL course, MHRD India

Bhadeshia H K D H Introduction to crystallography, University of Cambridge, course MP1

Callister W D 2007 *Callister's Materials Science and Engineering* (Indian Adaptation adapted by R Balasubramaniam) (New Delhi: Wiley)

Cullity B D 1978 *Elements of X-ray Diffraction* (Reading, MA: Addison Wesley)

Hammond C 2001 *The Basics of Crystallography and Diffraction* (Oxford: Oxford University Press)

Raghavan V 2004 *Materials Science and Engineering* 5th edn (Englewood Cliffs, NJ: Prentice-Hall)

Villars P and Calvert L D 1985 *Pearson's Handbook of Crystallographic Data for Intermetallic Phases* vol 1–3 (Metals Park, OH: American Society for Metals)

IOP Concise Physics

Concepts in Physical Metallurgy
Concise lecture notes
A Lavakumar

Chapter 3

Solidification

The three classic states of matter are gas, liquid and solid. In the gaseous state metal atoms occupy a greater amount of space due to their rapid motion. The atoms move independently and are usually widely separated, so the attractive forces between atoms are negligible. The arrangement of atoms in a gas is complete disorder. At a lower temperature, the kinetic energy of the atoms decreases and thus the attractive forces become large enough to bring most of the atoms together into a liquid. There is a continual interchange of atoms between the gas and the liquid across the liquid surface. The attractive forces between atoms in a liquid may be demonstrated by the application of pressure. A gas may be compressed into a smaller volume, but it takes high pressure to compress a liquid. There is, however, still enough free space in the liquid to allow the atoms to move about irregularly. When the temperature is decreased further, the motion becomes less vigorous and the attractive forces pull the atoms closer together until the liquid solidifies. Most materials contract upon solidification, indicating a closer packing of atoms in the solid state. The atoms in the solid are not stationary, but are vibrating around fixed points, giving rise to the orderly arrangement of crystal structures.

3.1 The mechanism of crystallization

Crystallization is the transition from liquid to the solid state and occurs in two stages:
- formation of nuclei and
- crystal growth.

Although the atoms in the liquid state do not have any definite arrangement, it is possible that some atoms at any given instant are in positions exactly corresponding to the space lattice they assume when solidified. These chance aggregates or groups are not permanent, but continually break up and reform at other points. The higher the temperature, the greater the kinetic energy of the atoms and the shorter the life of

doi:10.1088/978-1-6817-4473-5ch3

the group. When the temperature of the liquid is decreased the atom movement decreases, lengthening the life of the group, and more groups will be present at the same time. Atoms in a material have both kinetic and potential energy. *Kinetic energy* is related to the speed at which the atoms move and is strictly a *function of temperature.* The higher the temperature, the more active are the atoms and the greater is their kinetic energy. *Potential energy*, on the other hand, is related to the *distance between atoms.* The greater the average distance between atoms, the greater their potential energy.

3.2 The solidification of metals

A metal in a molten condition possesses high energy. As the melt cools, it loses energy to form crystals. Since heat loss is more rapid near mold walls than at any other place, the first submicroscopic metal crystals, called nuclei, form here. The melt experiences difficulty in starting to crystallize if no nuclei in the form of impurities are present to initiate the crystallization. However, under such conditions, the melt undercools and thus nuclei or seed crystals are formed. Thus-formed nuclei tend to grow at the second stage of solidification.

Crystal growth proceeds in three dimensions, the atoms attaching themselves in certain preferred directions, usually along the axes of the crystal. This gives rise to a characteristic tree-like structure which is called a dendrite. Since each nucleus is formed by chance, the crystal axes are pointed at random and the dendrites growing from them will grow in different directions in each crystal. Finally, as the amount of liquid decreases, the gaps between the arms of the dendrite will be filled and the growth of the dendrite will be mutually obstructed by that of its neighbors. This leads to a very irregular external shape. The crystals found in all commercial metals are commonly called grains because of this variation in external shape. This mismatch leads to a non-crystalline (amorphous) structure at the grain boundary with the atoms irregularly spaced. Since the last liquid to solidify is generally along the grain boundaries, there tends to be a higher concentration of impurity atoms in that area. Due to the chilling action of the mold wall, a thin skin of solid metal is formed at the wall surface immediately after pouring. The grain structure in a casting of a pure metal shows randomly oriented grains of small size near the mold wall and large columnar grains oriented toward the center of the casting.

3.2.1 Solidification of pure metals

Pure metals possess excellent thermal and electrical conductivity (e.g. Cu and Al), and higher ductility and better corrosion resistance compared to alloys. Pure metals melt and solidify at a single temperature, which may be termed as the melting point or freezing point (FP); above the FP the metal is liquid and below the FP it is in the solid state. If a number of temperature measurements are taken at different times while the pure metal is cooled under equilibrium conditions from the molten to solid states, a time–temperature plot (generally called a cooling curve) such as that in figure 3.1 will be obtained.

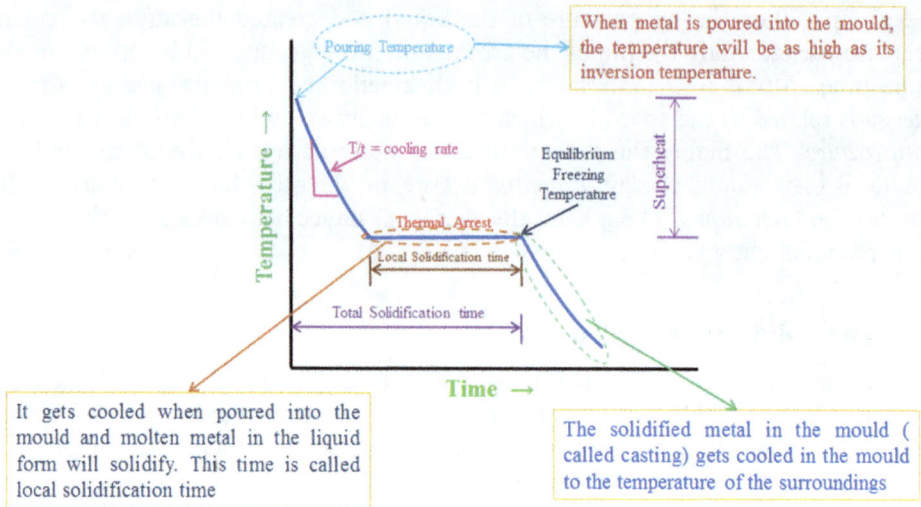

Figure 3.1. Cooling curve of pure metal.

3.2.2 The concept of supercooling

At the FP of a pure metal, both the liquid and solid states are at the same temperature. The kinetic energy of the atoms in the liquid and the solid must be the same, but there is a significant difference in potential energy. The atoms in the solid are much closer together, so solidification occurs with a release of energy. This difference in potential energy between the liquid and solid states is known as the *latent heat of fusion.* However, energy is required to establish a surface between the liquid and the solid. In pure materials, insufficient energy is released at the FP by the heat of fusion to create a stable boundary, and some undercooling is always necessary to form stable nuclei. Subsequent release of the heat of fusion will raise the temperature to the FP. The amount of undercooling required may be reduced by the presence of solid impurities, which reduce the amount of surface energy required. When the temperature of the liquid metal has dropped sufficiently below its FP, stable aggregates or nuclei appear spontaneously at various points in the liquid. These nuclei, which have now solidified, act as centers for further crystallization. As cooling continues, more atoms tend to freeze, and they may attach themselves to already existing nuclei or form new nuclei of their own. Each nucleus grows by the attraction of atoms from the liquid into its space lattice.

For better solidification of pure metals, cooling curves of the type in figure 3.2, i.e. with supercooling, are adopted. The liquid cools as the specific heat decreases (between points *A* and *B*). Undercooling is thus necessary (between points *B* and *C*). As the nucleation begins (point *C*), the latent heat of fusion is released, causing an increase in the temperature of the liquid. This process is known as recalescence (point *C* to point *D*). The metal continues to solidify at a constant temperature (T_{melting}). At point *E* solidification is complete. The solid casting continues to cool from this point.

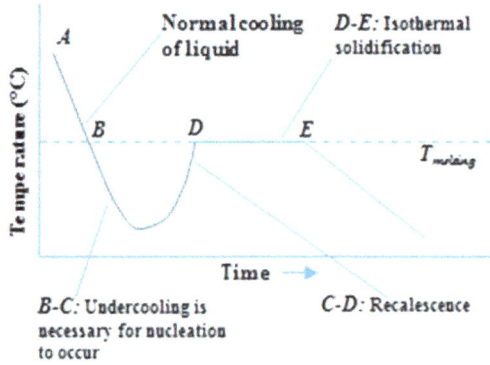

Figure 3.2. Cooling curve of pure metal involving supercooling.

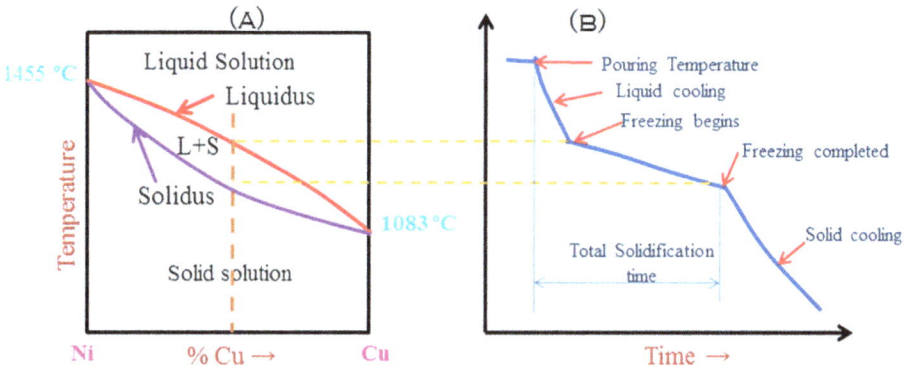

Figure 3.3. (a) A Ni–Cu phase diagram and (b) the cooling curve of a Ni–Cu alloy system.

3.2.3 The solidification of alloys

Pure metals are alloyed to improve their properties for specific applications, such as tensile strength, better corrosion resistance, improved machinability. Most alloys freeze over a temperature range (see figure 3.3(b)), i.e., they exhibit differential freezing characteristics. Differential freezing promotes growth in a manner other than by the advance of a smooth interface. Copper–nickel and gold–silver are two examples of solid solution alloys. Figure 3.3 (a) shows the phase diagram for a Ni–Cu alloy system and the cooling curve for a selected alloy system of Ni–Cu is shown in figure 3.3(b). Unlike in pure metals, solidification occurs throughout the temperature range. The latent heat of fusion is liberated gradually from B to C and this tends to increase the time required for solidification.

3.3 Nucleation

Nucleation can be defined as localized formation of a distinct thermodynamic phase. Nucleation can occur in the gas, liquid or solid phase. Some examples of phases that may form via nucleation include in-gas creation of liquid droplets in saturated vapor, in-liquid formation of gaseous bubble crystals (e.g. ice formation from water) or glassy regions, and in-solid nucleation of crystalline, amorphous and even vacancy clusters. Such solid-state nucleation is important, for example, to the semiconductor industry. Most nucleation processes are more physical than chemical.

There are two types of nucleation: homogeneous and heterogeneous. The distinction between them is made according to the site at which nucleating events occur. In homogeneous nucleation, the nuclei of the new phase form uniformly throughout the parent phase, whereas in heterogeneous nucleation, the nuclei form preferentially at structural inhomogeneities, such as container surfaces, the grain boundaries of insoluble impurities, dislocations and so on.

The probability of nucleation occurring at any point of the parent phase is the same throughout the parent phase. In heterogeneous nucleation there are some preferred sites in the parent phase where nucleation can occur. Compared to heterogeneous nucleation, (which starts at nucleation sites on surfaces) homogeneous nucleation occurs with much more difficulty in the interior of a uniform substance. The creation of a nucleus implies the formation of an interface at the boundaries of a new phase. Liquids cooled below the maximum heterogeneous nucleation temperature (melting temperature) but which are above the homogeneous nucleation temperature (pure substance freezing temperature) are supercooled. An example of supercooling is pure water, which freezes at −42 °C rather than the usual freezing temperature of water 0 °C.

There is no change in composition involved when we consider a pure metal. If we solidify an alloy this will involve long range diffusion. When a volume of material (V) transforms, three energies have to be considered: a reduction in G (the bulk Gibbs free energy, assuming we are working at constant T and P), an increase in γ (the interfacial free energy) and an increase in strain energy. In a liquid to solid phase transformation the strain energy term can be neglected as the liquid melt can flow to accommodate the volume change (assume we are working at constant T and P).

The process can start only below the melting point of the liquid (as only below the melting point is $G_{liquid} < G_{solid}$), i.e. we need to undercool the system. As we shall note, under suitable conditions (e.g. container-less solidification in zero gravity conditions), melts can be undercooled to a great extent without solidification taking place.

3.4 Homogeneous nucleation

Consider a pure melt, which is cooled below its melting temperature. Such a liquid is said to be undercooled.

Let a small volume V_s of the liquid turn into a solid. If G_s is the free energy per unit volume of the solid and G_l that of the liquid, if A_{ls} is the liquid–solid interfacial area and if γ_{ls} is interfacial free energy, then the total free energy of the system is given by

$$G_F = V_s G_s + V_l G_l + \gamma_{ls} A_{ls}.$$

Before the small volume of solid is formed, the free energy of the system is given by

$$G_I = (V_s + V_l)G_l.$$

Thus, the total free energy change associated with the transformation of a small volume of liquid into a solid is given by

$$\delta G = G_F - G_I$$
$$= V_s(G_s - G_l) + \gamma_{ls}A_{ls}$$
$$\delta G = V_s\Delta G + \gamma_{ls}A_{ls}$$

δG (free energy change on nucleation) = (volume) × (ΔG) + (surface) × (γ).

Note that the interfacial energy is always positive. Hence the contribution from the second term is always positive. However, depending on whether the liquid is above or below the melting temperature, ΔG is positive or negative. Hence, in an undercooled liquid, where ΔG is negative, the system will try to minimize the shape in such a way that the overall interfacial energy is reduced so that the maximum reduction in free energy can be achieved. If we assume the interfacial energy to be isotropic, then the volume transformed is a sphere of radius r (since the maximum volume is enclosed for the minimal surface area for a sphere), we obtain

$$\delta G = \left(\frac{4}{3}\pi r^3\right) \times (\Delta G_V) + (4\pi r^2) \times (\gamma_{ls}).$$

In figure 3.4 we show the variation of the volume free energy, interfacial energy and overall free energy change as a function of r. Since the interfacial contribution is r^2 and that of bulk free energy is r^3, at smaller r the interfacial energy always dominates and, being a positive energy, it actually suppresses the formation of the solid. Unless the size of the solid is above some size such that the (negative) bulk free energy change can more than compensate for the (positive) interfacial energy, the solid will not be stable (even if it forms). Thus, one can identify the critical radius of the solid (figure 3.5) that is stable when formed in the undercooled liquid by minimizing δG with respect to r:

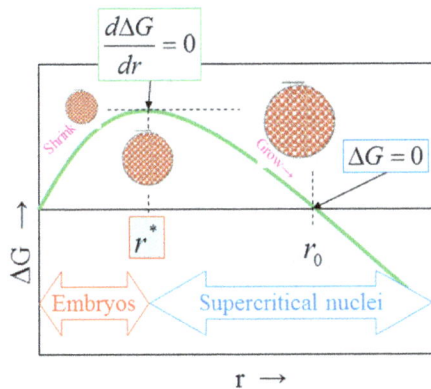

Figure 3.4. Variation of volume free energy with the radius of the nuclei formed. Reproduced with permission of Professor Subramaniam, Indian Institute of Technology http://home.iitk.ac.in/~anandh/E-book/.

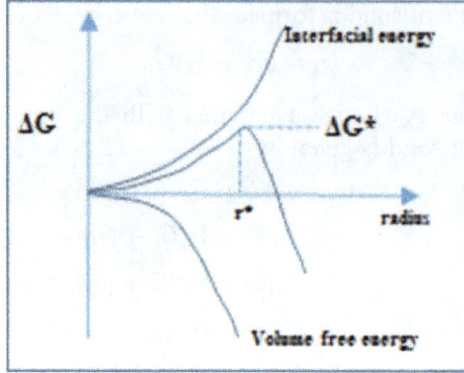

Figure 3.5. The relation between the interfacial energy and the volume free energy of a spherical nucleus.

$$\left[\frac{\mathrm{d}\delta G}{\mathrm{d}r}\right]_{r=r*} = 0$$

$$\delta G = \left(\frac{4}{3}\pi r^3\right) \times (\Delta G_V) + (4\pi r^2) \times (\gamma_{\mathrm{ls}})$$

$$\frac{\mathrm{d}}{\mathrm{d}r}\left[\frac{4}{3}\pi r^3 \Delta G + 4\pi r^2 \gamma_{\mathrm{ls}}\right]_{r=r*} = 0$$

$$[4\pi r^2 \Delta G + 8\pi r \gamma_{\mathrm{ls}}]_{r=r*} = 0$$

$$[4\pi r^2 \Delta G = -8\pi r \gamma_{\mathrm{ls}}]_{r=r*}$$

$$r* = \frac{-2\gamma_{\mathrm{ls}}}{\Delta G}$$

$$\delta G* = \frac{4}{3}\pi\left(\frac{-2\gamma_{\mathrm{ls}}}{\Delta G}\right)^3 \Delta G + 4\pi\left(\frac{-2\gamma_{\mathrm{ls}}}{\Delta G}\right)^2 \gamma_{\mathrm{ls}}$$

$$= -\frac{4}{3}\pi \times \frac{8\gamma_{\mathrm{ls}}^3}{\Delta G^3}\Delta G + 4\pi \times \frac{4\gamma_{\mathrm{ls}}^2}{\Delta G^2}\gamma_{\mathrm{ls}}$$

$$= \left(-\frac{32\pi}{3} + \frac{48\pi}{3}\right)\frac{\gamma_{\mathrm{ls}}^3}{\Delta G^2}$$

$$\delta G* = \frac{16\pi\gamma_{\mathrm{ls}}^3}{3\Delta G^2}.$$

What is the effect of undercooling (ΔT) on $r*$ and $\Delta G*$? We have noted that ΔG_V is a function of undercooling (ΔT). At larger undercooling ΔG_V increases and hence $r*$ and $\Delta G*$ decrease. This is evident from the equations for $r*$ and $\Delta G*$ as seen below (see also figure 3.6). At T_{m}, ΔG_V is zero and $r*$ is infinity, thus the melting point is not the same as the FP! This energy (G) barrier to nucleation is called the *nucleation barrier*.

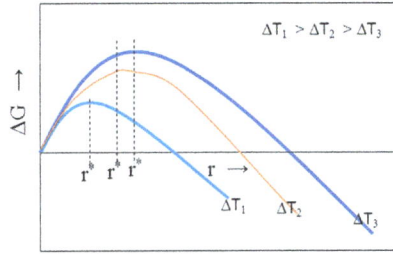

Figure 3.6. Variation of critical radius and driving force for nucleation w.r.t. undercooling.

The bulk free energy reduction is a function of undercooling

$$r^* = -\frac{2\gamma}{\Delta G_V} \quad \Delta G^* = \frac{16}{3}\pi\frac{\gamma^3}{\Delta G_V^2}.$$

Using the Turnbull approximation (linearizing the G–T curve close to T_m), we can obtain the value of ΔG in terms of the enthalpy of solidification:

$$\Delta G^* = \frac{16}{3}\pi\gamma^3\frac{T_m^2}{\Delta T^2 \Delta H^2}.$$

3.4.1 Homogenous nucleation rate

The process of nucleation (of a crystal from a liquid melt, below T_m^{bulk}) we have described so far is a dynamic one. Various atomic configurations are being formed in the liquid state, some of which resemble the stable crystalline order. Some of these *crystallites* are of a critical size $r^*_{\Delta T}$ for a given undercooling (ΔT). These crystallites can grow and become supercritical, thus transforming the melt into a solid. Crystallites smaller than r^* (embryos) tend to 'dissolve'. As the whole process is dynamic, we need to describe the process in terms of the rate, and the nucleation rate is $dN/dt \equiv$ number of nucleation events/time. True nucleation is the rate at which crystallites become supercritical. To find the nucleation rate, we have to find the number of critical-sized crystallites (N^*) and multiply it by the frequency/rate at which they become supercritical. If the total number of particles (which can act as potential nucleation sites, in homogenous nucleation for now) is N_t, then the number of critical-sized particles given by an Arrhenius type function with an activation barrier of ΔG^* is

$$N^* = N_t e^{\left(-\frac{\Delta G^*}{kT}\right)}.$$

The number of potential atoms which can jump to make the critical nucleus supercritical, the atoms which are 'adjacent' to the liquid, is s^*. If the lattice vibration frequency is ν and the activation barrier for an atom facing the nucleus (i.e. an atom belonging to s^*) to jump into the nucleus (to make in supercritical) is

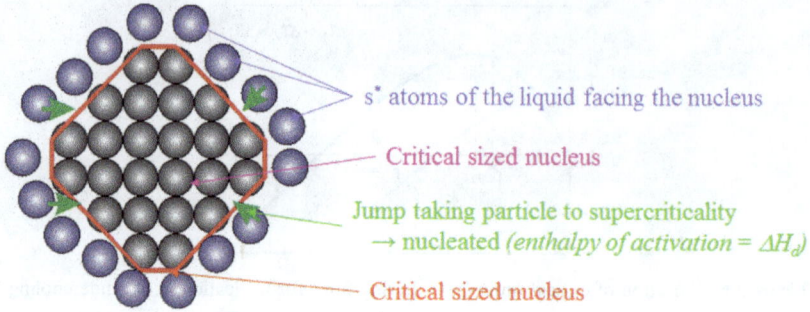

Figure 3.7. Formation of a critical nucleus. Reproduced with permission of Professor Subramaniam, Indian Institute of Technology http://home.iitk.ac.in/~anandh/E-book/.

ΔH_d, the frequency with which nuclei become supercritical due to atomic jumps into the nucleus is given by (see figure 3.7)

$$\nu' = S^*\nu e^{\left(-\frac{\Delta H_d}{kT}\right)}.$$

So the rate of nucleation is

Rate of nucleation = No. of critical sized particles

$\times$ Frequency with which they become supercritical

$$I = \frac{dN}{dt} = N_t e^{\left(-\frac{\Delta G^*}{kT}\right)} \times s^*\nu e^{\left(-\frac{\Delta H_d}{kT}\right)}$$

$$I = N_t s^*\nu e^{-\left(\frac{\Delta G^* + \Delta H_d}{kT}\right)}$$

3.5 Heterogeneous nucleation

Heterogeneous nucleation can be considered as a surface-catalyzed nucleation process. The extent of how a surface can catalyze or facilitate the nucleation depends on the contact angle of the nucleus with respect to the substrate (see figure 3.8). The smaller the angle (or the stronger the wetting on the surface), the lower the free energy change and the nucleation barrier will be. The critical radius of the nucleus (r^*) for heterogeneous nucleation is the same as that for homogeneous nucleation, whereas the critical volume of the nucleus (such as a droplet of liquid nucleated from the gas/vapor phase) is usually smaller for heterogeneous nucleation than for homogeneous nucleation, due to surface wetting (spreading). Heterogeneous nucleation occurs much more often than homogeneous nucleation. Heterogeneous nucleation applies to the phase transformation between any two phases of gas, liquid or solid; typically, for example, condensation of gas/vapor, solidification from liquid, bubble formation from liquid, etc. In the solidification of a liquid the nucleation site could be on the mold walls. For solid

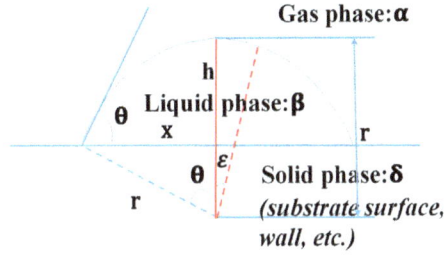

Figure 3.8. Heterogeneous nucleation.

state transformation suitable nucleation sites are non-equilibrium defects such as excess vacancies, dislocations, grain boundaries, stacking faults, inclusions and surfaces. One way to visualize the ease of heterogeneous nucleation is to consider that heterogeneous nucleation at a defect will lead to destruction/modification of the defect (making it less 'defective'). This will lead to some free energy ΔG_d being released and thus reducing the activation barrier. There are three different surface energies of interest:

$\gamma_{\alpha\beta}$ = surface of gas/liquid

$\gamma_{\beta\delta}$ = surface of liquid/solid

$\gamma_{\alpha\delta}$ = surface of gas/solid

$\Delta G = V\Delta G_V + A\gamma$

surface energy $= A_{\alpha\beta}\gamma_{\alpha\beta} + A_{\beta\delta}\gamma_{\beta\delta} + A_{\alpha\delta}\gamma_{\alpha\delta}$

$A_{\alpha\beta} = 2\pi rh \Rightarrow 2\pi r(r - r\cos\theta) \Rightarrow 2\pi r^2(1 - \cos\theta)$

$A_{\beta\delta} = \pi r^2 \Rightarrow \pi(r\sin\theta)^2 \Rightarrow \pi r^2 \sin^2\theta$

$A_{\alpha\delta} = \text{total area} - A_{\beta\delta} \Rightarrow A_o - \pi r^2 \sin^2\theta \; (\therefore A_o = \text{total area})$

surface energy $= \gamma_{\alpha\beta}2\pi r^2(1 - \cos\theta) + \gamma_{\beta\delta}\pi r^2 \sin^2\theta + \gamma_{\alpha\delta}A_o - \pi r^2 \sin^2\theta$

neglect the $\gamma_{\alpha\delta}A_o$ term, then the above equation becomes

surface energy $= \gamma_{\alpha\beta}2\pi r^2(1 - \cos\theta) + \gamma_{\beta\delta}\pi r^2 \sin^2\theta - \pi r^2 \sin^2\theta$

volume of spherical cap = area × height

$$\text{volume} = \int_0^\theta \pi(r\sin\varepsilon)^2 \times dH$$

We know that

$$H = r - r \cos \varepsilon \Rightarrow dH = r \sin \varepsilon d\varepsilon$$

$$= \int_0^\theta \pi r^2 \sin^2 \varepsilon \times r \sin \varepsilon d\varepsilon$$

$$= \int_0^\theta \pi r^3 \sin^2 \varepsilon \times \sin \varepsilon d\varepsilon$$

$$= \pi r^3 \int_0^\theta (1 - \cos^2 \varepsilon) \sin \varepsilon d\varepsilon$$

$$= \pi r^3 \int_0^\theta (\sin \varepsilon - \sin \varepsilon \cos^2 \varepsilon) d\varepsilon$$

$$= \pi r^3 \int_0^\theta \sin \varepsilon - \pi r^3 \int_0^\theta \sin \varepsilon \cos^2 \varepsilon d\varepsilon$$

Let us take

$$t = \cos \varepsilon \Rightarrow \frac{dt}{d\varepsilon} = -\sin \varepsilon \Rightarrow d\varepsilon = \frac{-dt}{\sin \varepsilon}$$

$$= \pi r^3 (-\cos \varepsilon)_0^\theta - \pi r^3 \int_0^\theta \sin \varepsilon t^2 - \frac{dt}{\sin \varepsilon}$$

$$= \pi r^3 (-\cos \varepsilon)_0^\theta + \pi r^3 \int_0^\theta t^2 dt$$

$$= \pi r^3 (-\cos \theta + 1) + \pi r^3 \left(\frac{t^3}{3} \right)_0^\theta$$

$$= \pi r^3 (1 - \cos \theta) + \frac{\pi r^3}{3} (\cos^3 \theta - \cos^3 0)$$

$$= \pi r^3 (1 - \cos \theta) + \frac{\pi r^3}{3} (\cos^3 \theta - 1)$$

$$= \pi r^3 \left[\frac{3 - 3 \cos \theta + \cos^3 \theta - 1}{3} \right]$$

$$= \frac{\pi r^3}{3} [2 - 3 \cos \theta + \cos^3 \theta]$$

The surface tension force balance is from figure 3.8

$$\cos\theta\gamma_{\alpha\beta} = \gamma_{\alpha\delta} - \gamma_{\beta\delta}$$

$$\cos\theta = \frac{\gamma_{\alpha\delta} - \gamma_{\beta\delta}}{\gamma_{\alpha\beta}}$$

$$\Delta G = \frac{\pi r^3}{3}[2 - 3\cos\theta + \cos^3\theta]\Delta G_V + \gamma_{\alpha\beta}2\pi r^2(1 - \cos\theta)$$

$$+ \pi r^2 \sin^2\theta(\gamma_{\beta\delta} - \gamma_{\alpha\delta})$$

put $\gamma_{\beta\delta} - \gamma_{\alpha\delta} = -\gamma_{\alpha\beta}\cos\theta$

$$= \frac{\pi r^3}{3}[2 - 3\cos\theta + \cos^3\theta]\Delta G_V + \gamma_{\alpha\beta}2\pi r^2(1 - \cos\theta) - \gamma_{\alpha\beta}\pi r^2\sin^2\theta\cos\theta$$

$$= \frac{4\pi r^3}{3}\frac{[2 - 3\cos\theta + \cos^3\theta]}{4}\Delta G_V + \pi r^2\gamma_{\alpha\beta}(2 - 2\cos\theta - \sin^2\cos\theta)$$

$$= \frac{4\pi r^3}{3}\frac{[2 - 3\cos\theta + \cos^3\theta]}{4}\Delta G_V$$

$$+ \pi r^2\gamma_{\alpha\beta}(2 - 2\cos\theta - (1 - \cos^2\theta)\cos\theta)$$

$$= \frac{4\pi r^3}{3}\frac{[2 - 3\cos\theta + \cos^3\theta]}{4}\Delta G_V + 4\pi r^2\gamma_{\alpha\beta}\frac{[2 - 3\cos\theta + \cos^3\theta]}{4}$$

$$= \frac{[2 - 3\cos\theta + \cos^3\theta]}{4}\left[\frac{4\pi r^3}{3}\Delta G_V + 4\pi r^2\gamma_{\alpha\beta}\right]$$

$$= \left[\frac{4\pi r^3}{3}\Delta G_V + 4\pi r^2\gamma_{\alpha\beta}\right]f(\theta) \rightarrow \text{where } f(\theta) = \frac{[2 - 3\cos\theta + \cos^3\theta]}{4}$$

$$\Delta G_{\text{hetero}} = \Delta G_{\text{homo}}f(\theta)$$

$$\Delta G^*_{\text{hetero}} = \Delta G_{\text{hetero}}(r^*) = \Delta G_{\text{homo}}(r^*)f(\theta) = \Delta G^*_{\text{homo}}f(\theta)$$

The nucleation barrier can be significantly lower for heterogeneous nucleation due to the wetting angle affecting the shape of the nucleus. Using the procedure as before (for the case of homogeneous nucleation) we can find r^* for heterogeneous nucleation. Using the surface tension balance equation, we can write the formulae for r^* and ΔG^* using a single interfacial energy $\gamma_{\alpha\beta}$ (and contact angle θ). Further, we can write $\Delta G^*_{\text{hetero}}$ in terms of ΔG^*_{homo} and contact angle θ (see figure 3.9).

3.5.1 Heterogeneous nucleation rate

The rate of heterogeneous nucleation can be expressed in a form similar to that of the homogeneous nucleation rate. In addition to the difference in the ΔG^* term and

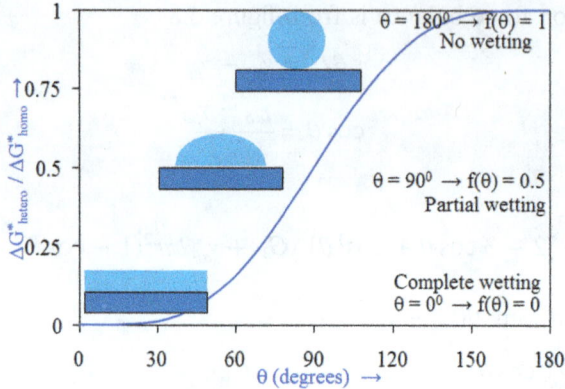

Figure 3.9. ΔG^*_{hetero} in terms of ΔG^*_{homo} and contact angle θ.

the pre-exponential term, we will include only the number of preferred nucleation sites, which is many orders of magnitude smaller than the number of atoms per unit volume used in the homogeneous case. In order of magnitude, the number of nucleation sites for various situations is typically as follows.

If N_i is the number of sites per unit volume having a nucleation barrier ΔG_i^*, then we can write the expression for the heterogeneous nucleation rate as

$$I_{hetero} = N_i s_i^* \nu e^{-\left(\frac{\Delta G^* + \Delta G_d}{RT}\right)}.$$

Usually, $I_{hetero} > I_{homo}$, because $\Delta G^*_{hetero} < \Delta G^*_{homo}$ dominates the result through the pre-exponential factor, which can be much smaller for heterogeneous nucleation compared to the homogeneous case.

$$I = \frac{dN}{dt} = N_t e^{\left(-\frac{\Delta G^*}{kT}\right)} \times s^* \nu e^{\left(-\frac{\Delta H_d}{kT}\right)}$$

$$I = N_t s^* \nu e^{-\left(\frac{\Delta G^* + \Delta H_d}{kT}\right)}$$

Further reading

Askeland D R and Phulé P 2006 *The Science and Engineering of Materials* (Boston, MA: Cengage Learning)

Avner S H 1997 *Introduction to Physical Metallurgy* (New York: McGraw-Hill)

Callister W D 2007 *Callister's Materials Science and Engineering* (Indian Adaptation adapted by R Balasubramaniam) (New Delhi: Wiley)

Raghavan V 2004 *Materials Science and Engineering* 5th edn (Englewood Cliffs, NJ: Prentice-Hall)

Subramaniam A and Balani K (IITK) *Materials Science and Engineering* (e-book) MHRD, India

Concepts in Physical Metallurgy
Concise lecture notes
A Lavakumar

Chapter 4

Crystal imperfections

The lattice structures and type of atomic arrangement in crystalline solids were introduced in previous chapters. The behavior of electrons determines the way in which the atoms interact and the type of bonding (metallic, ionic, covalent or van der Waals) that holds the atoms in a solid together. But is the knowledge of the bonding and crystal structure sufficient to predict the macroscopic properties of materials? So far in this book, for crystalline solids we had assumed a correspondence between the abstract three-dimensional lattice and the actual structure of solids. This implies that crystals are perfect.

A perfect crystal: a crystal in which all the atoms are at rest on their correct lattice position in the crystal. Such perfect crystals could only exist at a temperature of absolute zero, and thermal vibrations can be treated as a form of defect in crystal structures. For $T > 0$ K, defects always exist in the structure, i.e. *real crystals are never perfect, there are always defects*.

For example, if a bulk metal (e.g. steel) were a single perfect crystal, it would have a strength far exceeding the strongest steel ever produced from metallurgical research. This can be seen as a limitation, as the strength of metal comes from the number and type of defects, rather than from the nature of the ideal crystal lattice. As another example, if we buy a diamond ring, it is mostly the number and type of defects in the diamond crystal that define the amount of money we pay for a given diamond crystal. Many other important properties of materials are due to the imperfections caused by crystal defects. In this chapter, we will discuss different types of imperfections or defects in the ideal arrangement of atoms in a crystal.

4.1 Thermodynamic causes of crystal imperfections

Crystal defects are thermodynamically controlled phenomena. The origin of defects can be understood from the following example.

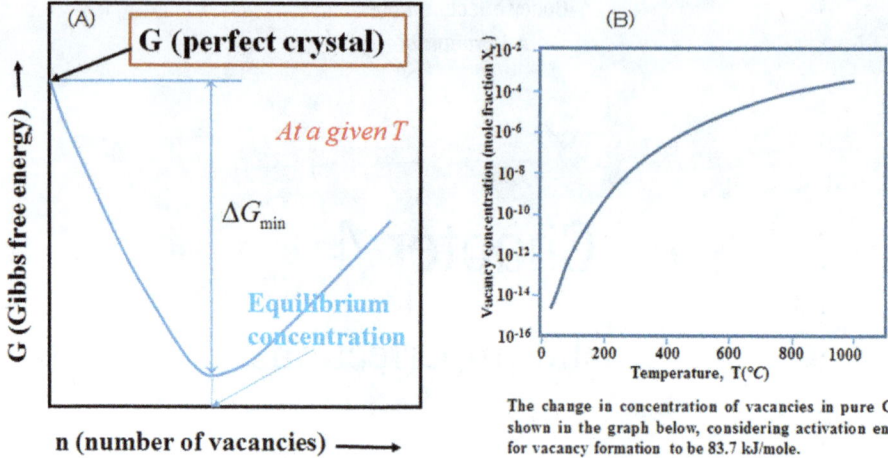

Figure 4.1. Variation of free energy with the number of vacancies. Note that the free energy for the system is minimum at a certain number of vacancies. This is called the equilibrium number of vacancies. (b) The number of vacancies present per mole of atoms at different temperatures.

Consider a single crystal containing about one mole of atoms. The effects of adding defects to the crystals can be explained using the two cases below. In case one, if we add one defect to the case of the single crystal, then the probability of locating the defect within the crystal is 10^{23}. Therefore, the crystal shows a large increase of entropy. In the second case, when we introduce one defect to the crystal containing already 10% defect then the probability of finding one defect is only 10. Thus only a small increase in entropy occurs.

This kind of entropy is termed as configurational entropy. This arises due to the arrangement of atoms and defects within the crystal. The equation giving the configurational entropy is

$$S = k \ln(\omega)$$

Where K is the Boltzmann constant and ω is the probability.

So far we have looked at the effect of entropy; however, the creation of any kind of defect costs energy, since the total lattice energy will be reduced. The competing effects of the energy required to disturb the lattice and the initial large gain on entropy cause a minimum in free energy. Therefore, at this concentration of vacancies the crystal becomes most stable and the concentration of vacancies is equilibrium concentration of defects. The variation of free energy with concentration of vacancies is shown in figure 4.1. The concentration of vacancies is dependent on the temperature. The variation of vacancy concentration with temperature for a copper crystal is given in figure 4.1(b).

4.2 Classification of crystal imperfections

Crystal imperfections can be divided into four basic types based on type of defect and its dimensionality. They are point defects (0-dimensional), line defects

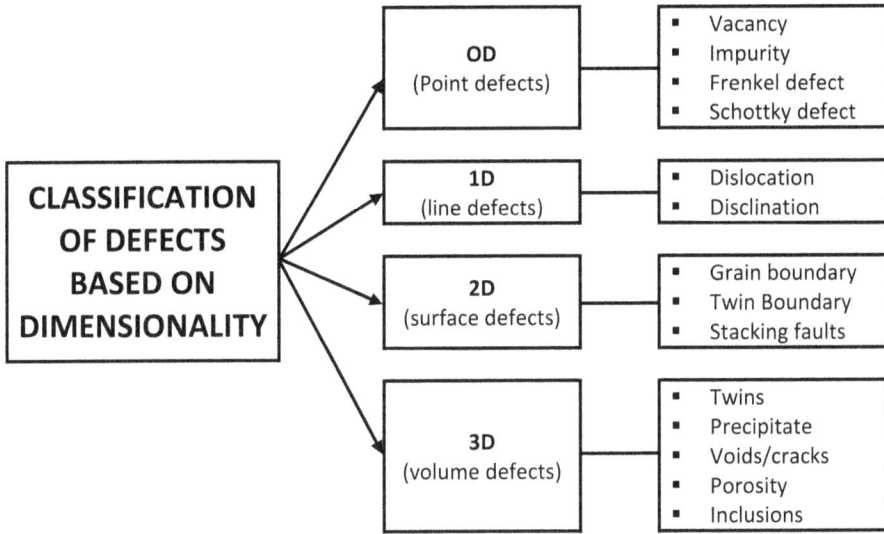

| OD (Point defects) | <ul><li>Vacancy</li><li>Impurity</li><li>Frenkel defect</li><li>Schottky defect</li></ul> |

Figure 4.2. Classification of crystal imperfections.

(1-dimensional), surface defects (2-dimensional) and volume defects (3-dimensional). Figure 4.2 shows a broad classification system for crystal imperfections.

4.3 Point defects

In a crystal lattice, a point defect is one which is completely local in its effect, e.g. a vacant lattice site. The introduction of a point defect into a crystal increases its internal energy compared to that of the perfect crystal.

The number of defects at equilibrium at a certain temperature can be determined from

$$n_d = Ne^{\frac{-E_d}{kT}}$$

where n_d is the number of defects, N is the total number of atomic sites per cubic meter or per mole, E_d is the energy of activation necessary to form the defects, k is the Boltzmann constant and T is the absolute temperature. The possible point defects are described in the following sections.

4.3.1 Vacancies

A vacancy or vacant site implies an unoccupied atom position within a crystal lattice. In other words, vacancies are simply empty atom sites (see figure 4.3(a)). It can be shown through thermodynamic reasoning that lattice vacancies are a stable feature of metals at all temperatures above absolute zero. The everyday industrial processes of annealing, homogenization, precipitation, sintering, surface hardening, oxidation and creep all involve, to varying degrees, the transport of atoms through the lattice with the help of vacancies. Vacancies exist in a certain proportion in a crystal in thermal equilibrium, leading to an increase in the randomness of the structure.

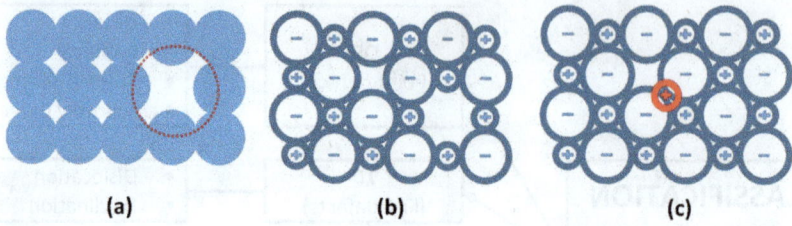

Figure 4.3. (a) Vacancies created in a lattice, (b) a Schottky defect and (c) a Frankel defect.

Vacancies may occur as a result of imperfect packing during the original crystallization or they may arise from thermal vibrations of atoms at elevated temperatures, because as thermal energy is increased there is a higher probability that individual atoms will jump out of their position of lowest energy. The atoms surrounding a vacancy tend to be closer together, thereby distorting the lattice planes. Vacancies may be single, or two or more of them may condense into a di-vacancy or tri-vacancy state. A Schottky defect is closely related to vacancies and is formed when an anion–cation pair in an ionic compound is displaced from its site and creates a vacancy of the charged pair (see figure 4.3(b)). In the case of Frankel defects in ionic solids, the cation, being smaller in size, can migrate to another lattice position, particularly to the interstitial sites inside the crystal lattice (see figure 4.3(c)). However, in both defects the overall electrical neutrality of the atoms is maintained.

4.3.2 Interstitials

In interstitials, atoms occupy positions between the atoms of the ideal crystal. The interstitial may be either an atom from the same crystal or a foreign atom. The interstitial atom may be lodged within a crystal structure, particularly if the APF is low. Interstitiality produces atomic distortion because the interstitial atom tends to push the surrounding atoms farther apart, unless the interstitial atom is smaller than the rest of the atoms in the crystal.

4.3.3 Impurities

Impurities give rise to compositional defects. Impurities may be small particles embedded in the structure, or foreign atoms in the lattice. Foreign atoms generally have atomic radii and electronic structures differing from those of the host atoms and, therefore, act as centers of distortion. Impurity atoms are introduced into crystal structures as substitutional or interstitial atoms (see figure 4.4). Foreign atoms either occupy lattice sites from which the regular atoms are missing or they occupy positions between the atoms of the ideal crystal. Impurities may considerably distort the lattice. Impurity defects occur in metallic, covalent and ionic solids and play a very important role in many solid state process such as diffusion, phase transformation, and electrical and thermal conductivity. The controlled addition of impurities to a very pure crystal is the basis of producing many electronic devices.

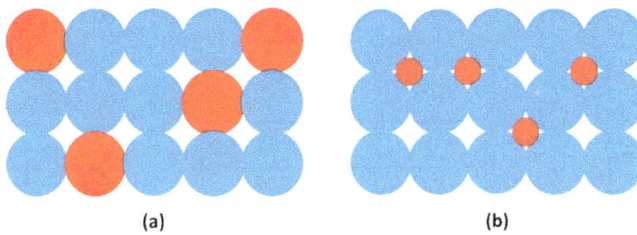

Figure 4.4. (a) Substitutional solid solution and (b) interstitial solid solution.

4.3.4 Electronic defects

Electronic defects are the result of errors of charge distribution in solids. These defects are free to move in the crystal under the influence of an electrical field, thereby accounting for some of the electronic conductivity of certain solids and their increased reactivity. A vacancy or an interstitial impurity may produce an excess or deficit of positive or negative charges through an excess of metal ions as in ZnO, where there is an excess of interstitial zinc zones.

4.4 Line defects

The most important two-dimensional or line defect is a dislocation. A dislocation may be defined as a disturbed region between two substantially perfect parts of a crystal. Dislocation is a defect in a crystal structure whereby a part-plane of atoms is displaced from its symmetrically stable position in the array. It is surrounded within the structure by an extensive elastic strain field and its associated stresses (see figure 4.5(a)). The dislocation is responsible for the phenomenon of slip, by which most metals plastically deform. One may conclude that the dislocation is the region of localized lattice disturbance separating the slipped and unslipped regions of a crystal (see figure 4.5(b)). Dislocations are intimately connected with slip and with many other mechanical phenomena such as strain hardening, the yield point, creep, fatigue and brittle fracture.

The two basic types of dislocations are:
- edge dislocation and
- screw dislocation.

The concepts behind and main difference between an edge dislocation and a screw dislocation are as follows. In an edge dislocation the Burger's vector lies at a right angle to the line of the dislocation, along the axis of rows of atoms in the same plane, whereas in screw dislocation the Burger's vector lies parallel to the dislocation line along the axis of a line of atoms in the same plane. The Burger's vector marks the magnitude and direction of the strain component of dislocation and can completely describe the dislocation if the orientation of the dislocation line is known.

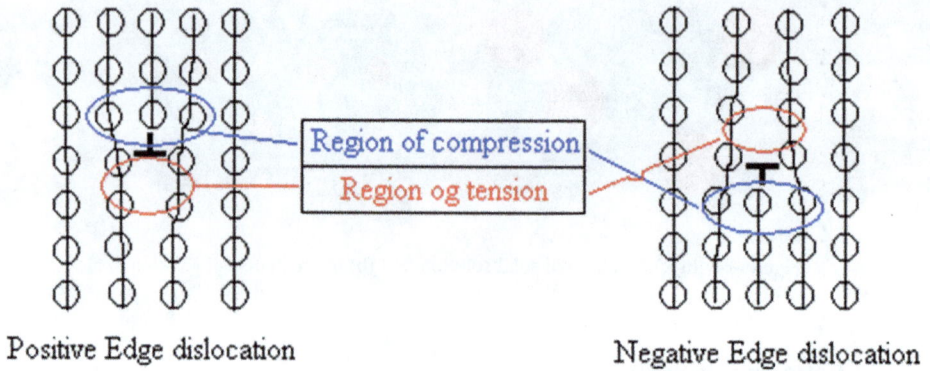

Positive Edge dislocation   Negative Edge dislocation

Figure 4.5. (a) Stress field distribution in the half planes of an edge dislocation and (b) a schematic diagram showing the slipped and unslipped part of a crystal system with a movable dislocation.

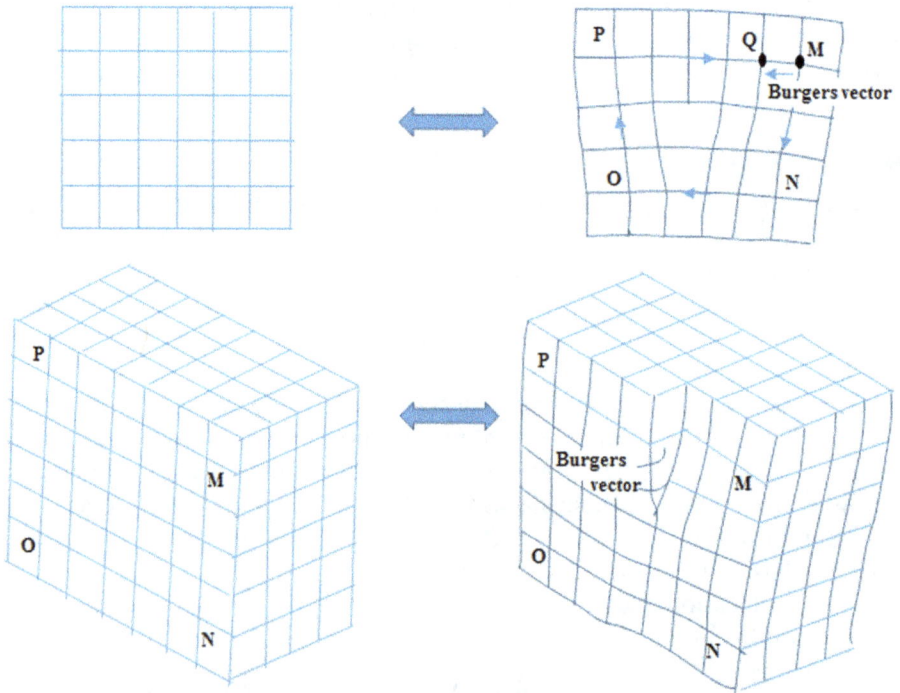

Figure 4.6. Edge and screw dislocations with Burger's vectors. Image credit: Javier Bartolomé Vílchez /CC-BY-SA-3.0 (https://creativecommons.org/licenses/by-sa/3.0/deed.en).

4.4.1 Edge dislocation

An edge dislocation lies perpendicular to its Burger's vector (see figure 4.6) and moves in the direction of the Burger's vector. An edge dislocation involves an extra row of atoms, either above or below the slip plane. The presence of this extra row means that adjacent atoms are displaced elastically, and consequently elastic forces are exerted on the dislocation from both sides. These forces balance out, so that it is easy to move the dislocation from one position to another. Under a shear stress sense ↔, a positive dislocation moves to the right and a negative dislocation to the left. The edge dislocation is particularly useful in explaining slip in plastic flow during mechanical working.

4.4.2 Screw dislocation

A screw dislocation lies parallel to its Burger's vector (see figure 4.6) and moves in a direction perpendicular to Burger's vector. In the screw dislocation, the distortion follows a helical or screw path, and both right-hand and left-hand paths are possible. A screw dislocation is a continuous helicoidal plane of atoms rather than a series of parallel planes. The forces required to form and move a screw dislocation, although probably somewhat greater than those required to initiate an edge dislocation, are markedly less than those required to exceed the elastic limit of a crystal. The speed of movement of a screw dislocation is less than that of an edge dislocation. Screw dislocations are particularly useful in explaining crystal growth as well as slip in plastic deformation. A detailed table is given to explain the difference between edge and screw dislocation (table 4.1).

Table 4.1. Differences between edge dislocation and screw dislocation.

Edge dislocation	Screw dislocation
(1) It may originate when there is a slight mismatch in the orientation of the adjacent part of the growing crystal so that an extra row of atoms is introduced or eliminated.	(1) It may originate during crystallization, when there is a twisting in stacking sequences of atoms and unit cell formation of a step around a line known as screw axis.
(2) The region of disturbance of a lattice extends along an edge inside the crystal.	(2) The region of disturbance of a lattice extends in two separate planes, which are perpendicular to each other.
(3) The edge dislocation has an incomplete plane, which lies above or below the slip plane.	(3) The screw dislocation has lattice planes which spiral around the dislocation like a left-hand or right-hand screw.
(4) In an edge dislocation, all the three types of stresses i.e. tensile, compressive and shear, may exist.	(4) In a screw dislocation, only shear stress may exist.
(5) A pure edge dislocation can glide or slip in a direction perpendicular to its length. However, it may move vertically by a process known as climb, if diffusion of atoms or vacancies can take place at an appreciable rate.	(5) The screw dislocation can move by slip. The movement by climb is not possible.
(6) The Burgers vector always lies perpendicular to an edge dislocation.	(6) The Burgers vector always lies parallel to the screw dislocation.

4.5 Surface defects

Surface defects may include grain boundaries, tilt boundaries, twin boundaries, stacking faults, coherent and incoherent precipitate interfaces.

4.5.1 Grain boundaries

Grain boundaries are those planar imperfect ions in polycrystalline materials that separate crystals of different orientation. A grain boundary is formed when two growing grain surfaces meet. In grain boundaries the atomic packing is imperfect. At the grain boundary between two adjacent grains there is a transition zone, which is not aligned with either grain. Most atoms at the boundaries are located in highly strained and distorted positions and their free energy is higher than that of the atoms in regular, undisturbed parts of the crystal lattice. Although a grain boundary thickness of a few atomic diameters has been confirmed using field ion micrography, boundaries between grains in commercial metals are often wider because impurities commonly concentrate at the grain boundaries during solidifications.

4.5.2 Tilt boundaries

A tilt boundary is another surface imperfection, and it may be regarded as an array of edge dislocations. In fact, tilt boundaries are a class of low-angle boundaries. By rotation of an axis in the boundary it is possible to bring the axis of two bordering grains into coincidence, i.e. a tilt boundary.

Twist boundaries, the second class of low-angle boundaries, result from a set of screw dislocations. In a twist boundary, the rotation is about an axis normal to the boundary. If the misorientation angle (θ), between the grains is greater than $10°$–$15°$, the resultant grain boundaries are called high-angle boundaries.

4.5.3 Twin boundaries

Other surface imperfections are twin boundaries. A twin boundary separates two parts of a crystal with the same orientation and which look like mirror images of each other.

4.5.4 Stacking faults

Stacking faults may arise where there is only a small dissimilarity between the stacking sequences of close-packed planes in fcc and hcp metals, e.g. ABCABC and ABABAB. It is possible for one atom layer to be out of sequence relative to the atoms of the layer above and below, giving a mistake, e.g. ABCACAB.

In other words, the stacking is a discrepancy in the packing sequence of the layers of atoms. A stacking fault exists when a metal crystal, as a result of a slip or another cause, has two adjacent crystallographic planes located together out of the normal geometric pattern within the system. Stacking faults are frequently observed in deformed metals than annealed metals.

4.6 Volume defects

Volume defects in crystals are three-dimensional aggregates of atoms or vacancies. Examples of volume defects are precipitates, inclusions and voids/pores, etc.

4.6.1 Precipitates

The precipitates are small particles that are introduced into the matrix by solid state reactions. While precipitates are used for several purposes, their most common purpose is to increase the strength of structural alloys by acting as obstacles to the motion of dislocations. Their efficiency in doing this depends on their size, their internal properties and their distribution through the lattice. However, their role in the microstructure is to modify the behavior of the matrix rather than to act as separate phases.

4.6.2 Inclusions

Inclusions are foreign particles or large precipitate particles. They are usually undesirable constituents in the microstructure. For example, inclusions have a deleterious effect on the useful strength of structural alloys since they are preferred sites for failure. They are also often harmful in microelectronic devices since they disturb the geometry of the device by interfering in manufacturing, or alter its electrical properties by introducing undesirable properties of their own.

4.6.3 Voids

Voids (or pores) are generally caused by gases that are trapped during solidification or by vacancy condensation in the solid state. They are almost always undesirable defects. Their principal effect is to decrease mechanical strength and promote fracture at small loads.

Further reading

ASM International 1990 *Irons and Steels and High Performance Alloys: Specialty Steels and Heat Resistant Alloys* (*Metals Handbook* vol 1) 10th edn (Materials Park, OH: ASM International) pp 755–1003

Callister W D 2007 *Callister's Materials Science and Engineering* (Indian Adaptation adapted by R Balasubramaniam) (New Delhi: Wiley)

Hirth J P and Lothe J *Theory of Dislocations* (New York: McGraw-Hill)

Honeycombe R W K 1984 *Plastic Deformation of Metals* (London: Edward Arnold)

Kelly A, Groves G W and Kidd P 2000 *Crystallography and Crystal Defects* (Chichester: Wiley)

Khurmi R S and Sedha R S 2000 *Material Science* (New Delhi: S Chand and Company Limited)

Kingery W D, Bowen H K and Uhlmann D R 1976 *Introduction to Ceramics* (New York: Wiley) ch 14

Raghavan V 2004 *Materials Science and Engineering* 5th edn (Englewood Cliffs, NJ: Prentice-Hall)

Royal Society (Great Britain), Metals Society. Metal Science Committee 1985 *Dislocations and Properties of Real Materials* (London: Institute of Metals)

Subramaniam A and Balani K (IITK) *Materials Science and Engineering* (e-book) MHRD, India

Chapter 5

Mechanical properties of materials

The term property, in a broader sense, may be defined as the quality which defines the specific characteristics of a metal. The mechanical properties of a material reflect the relationship between its response or deformation to an applied load or force. Important mechanical properties are strength, hardness, ductility and stiffness. These properties are ascertained by performing carefully designed laboratory experiments that replicate as closely as possible the service conditions. Factors to be considered include the nature of the applied load and its duration, as well as the environmental conditions. It is possible for the load to be tensile, compressive or shear, and its magnitude may be constant with time or it may fluctuate continuously. The application time may be only a fraction of a second or it may extend over a period of many years. Service temperature can also be an important factor.

5.1 Types of mechanical properties

The mechanical properties of a metal are those properties which completely define its behavior under the action of external loads or forces. Or, in other words, mechanical properties are those properties which are associated with its ability to resist failure and other behavior under the action of external forces. A sound knowledge of these properties is essential for an engineer to enable him/her to select a suitable metal for various structures or various components of a machine. Most mechanical properties of metals are generally expressed in terms of stress, strain or both. Although there are many mechanical properties of metals that an engineer should know, the following are important from our current point of view: elasticity, plasticity, ductility, brittleness, malleability, weldability, castability, hardness, toughness, stiffness, resilience, creep, endurance and strength.

5.1.1 Elasticity

The term elasticity may be defined as the property of the metal by virtue of which it is able to retain its original shape and size after the removal of the load. In nature, no

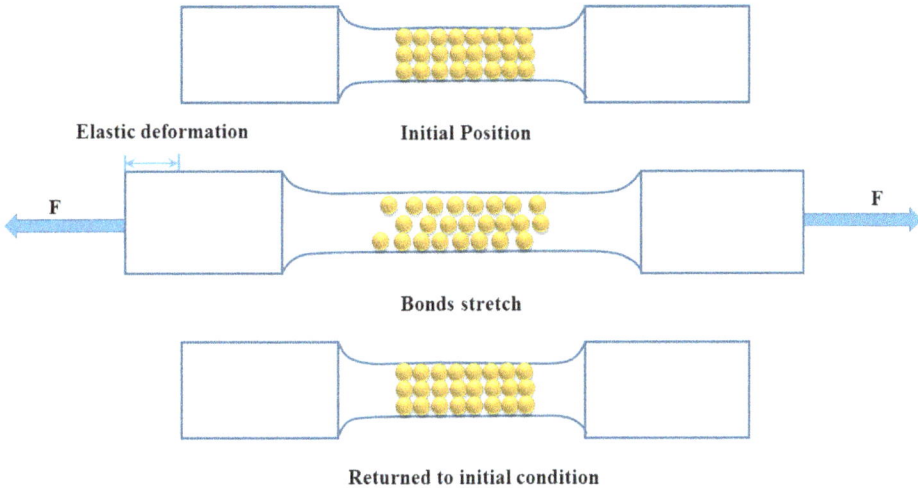

Elastic deformation   **Initial Position**

F   F

Bonds stretch

Returned to initial condition

Figure 5.1. Deformation of elastic material by stretching. It can be observed that the atoms return to their original positions after removal of the force.

material is perfectly elastic over the entire range of stress up to rupture. Steel and some other metals have a wide range over which they appear to be elastic. A schematic diagram showing the conditions of atoms during elastic deformation is given in figure 5.1.

Stress is the internal resistance set up in a material under the action of external forces. Mathematically, stress is expressed as force divided by cross-sectional area.

Strain is the deformation per unit length under the action of external forces. Mathematically, strain is expressed as the change in length divided by the original length. The strain can be classified into elastic strain and plastic strain. Elastic strain is a temporary strain, which appears as long as the external force is applied. It disappears after removal of the forces. Plastic strain is a permanent strain caused by external forces, when the stress exceeds the elastic limit. Plastic strain takes place as a result of permanent displacement of the atoms inside the material.

Elasticity is always desirable in metals used in machine tools and other structural members.

5.1.2 Plasticity

The term plasticity can be defined as the property of a metal by virtue of which a permanent deformation (without fracture) takes place (see figure 5.2), whenever it is subjected to the action of external forces. The plasticity of a metal depends upon its nature and the environmental conditions, i.e. whether the metal is shaped when red hot or cold.

Most metals have been found to possess good plasticity even at room temperature, while cast iron does not possess any appreciable plasticity even in red hot condition. This property has uses in forming, shaping and extruding operations of metals.

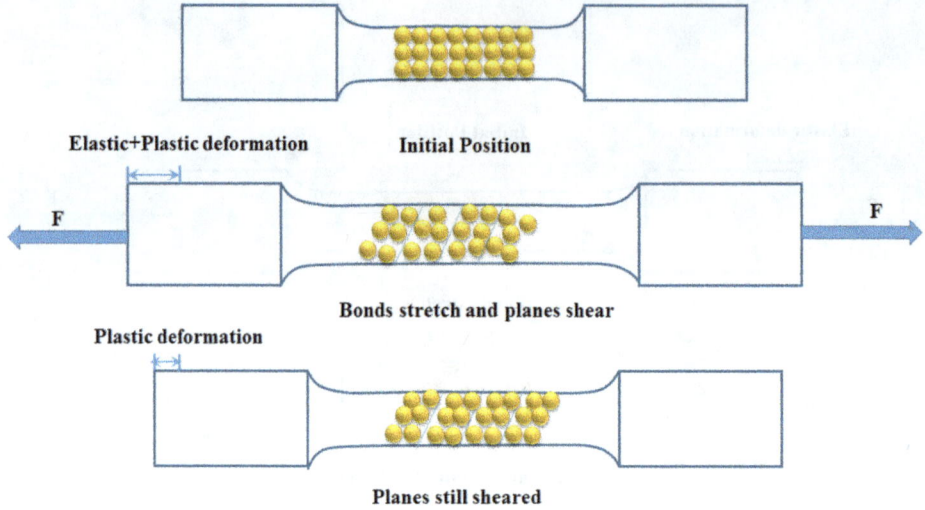

Figure 5.2. Plastic deformation in materials.

5.1.3 Ductility

The term ductility may be defined as the property of the metal by virtue of which it can be drawn into wires or elongated before rupture takes place. It depends upon the grain size of the metal crystals. The measures of the ductility of a metal are its percentage elongation and percentage reduction in the cross-sectional area before rupture. The term percentage elongation is the maximum elongation in the length expressed as a percentage of original length. Mathematically, percentage elongation is

$$= \frac{\text{increase in length}}{\text{original length}} \times 100.$$

Similarly, the term *percentage reduction of cross-sectional area* is the maximum decrease in cross-sectional area, expressed as the percentage of the original cross-sectional area. Mathematically, the percentage reduction of cross-sectional area is

$$= \frac{\text{decrease in cross-sectional area}}{\text{original cross-sectional area}} \times 100.$$

A little consideration will show that a metal with good percentage of elongation or reduction in cross-sectional area explains its high quality. Metals with more than 15% elongation are considered as ductile, metals with 5–15% elongation are considered to be of intermediate ductility, and metals with less than 5% elongation are considered to be brittle. A comparison between the ductility of two materials in terms of percentage of elongation is given in figure 5.3.

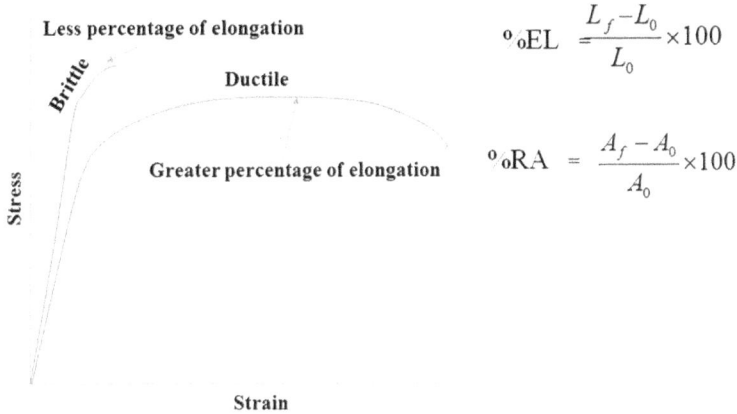

The diagram shows:

Less percentage of elongation

Brittle

Ductile

Stress (vertical axis)

Greater percentage of elongation

$$\%EL = \frac{L_f - L_0}{L_0} \times 100$$

$$\%RA = \frac{A_f - A_0}{A_0} \times 100$$

Strain

Figure 5.3. Engineering stress–strain diagram comparing the percentage of elongation in two materials.

5.1.4 Brittleness

The term brittleness may be defined as the property of a metal by virtue of which it will fracture without any appreciable deformation. The property is opposite to the ductility of a metal. Cast iron, glass and concrete are the examples of brittle materials. This property is important in the design of machine tools, which are subjected to sudden loads.

5.1.5 Hardness

The term hardness may be defined as the property of a metal by virtue of which it is able to resist abrasion, indentation (or penetration) and scratching by harder bodies. It is measured by the resistance the metal offers to scratching.

5.1.6 Toughness

The term toughness may be defined as the property of a metal by virtue of which it can absorb maximum energy before fracture takes place. The tenacity and hardness of a metal are measures of its toughness. It has been found that the value of toughness falls with an increase in temperature.

The importance of toughness is in the selection of a material where the load increases beyond the elastic limit or yield point, e.g. a power press punch or pneumatic hammer, etc. Toughness can be calculated as the area under the stress strain diagram (see figure 5.4).

5.1.7 Stiffness

The term stiffness may be defined as the property of a metal by virtue of which it resists deformation. The modulus of elasticity (i.e. the ratio of stress to the strain below the elastic limit) is a measure of the stiffness of a metal. The stiffness of a metal is of great importance when selecting it as a component of a machine or a structure. Stiffness is also used in graduating spring balances and spring controlled measuring instruments.

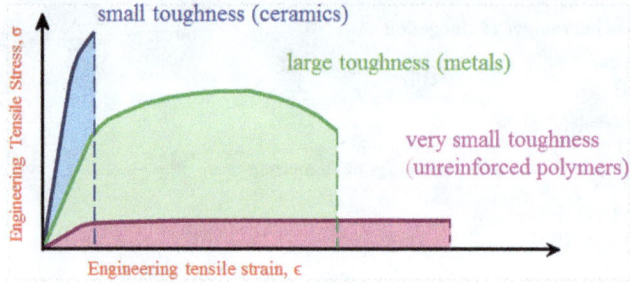

Figure 5.4. Comparison of the toughness of different materials.

5.1.8 Resilience

The term resilience may be defined as the property of a metal by virtue of which it stores energy and resists shocks and impacts. The resilience of a metal is measured by the amount of energy that can be stored, per unit volume, after it is stressed up to the elastic limit.

5.1.9 Creep

The term creep may be defined as the property of a metal by virtue of which it deforms continuously under a steady load. Generally, creep occurs in steel at higher temperatures. Creep is always considered when designing IC engines, boilers, turbines, etc. Creep is of great importance in the following cases:

- Soft metals used at about room temperature, such as lead pipes, white metal bearings, etc.
- Steam and chemical plants operating at 450–550 °C.
- Gas turbines working at higher temperatures.
- Rockets, missiles and supersonic jets.
- Nuclear reactor fields.

5.1.10 Endurance

The term endurance may be defined as the property of a metal by virtue of which it can withstand varying stresses. The maximum value of stress that can be applied for an indefinite time without causing failure is known as the endurance limit. It has been found that the endurance limit for ordinary steel is approximately half the tensile strength.

The endurance of a metal is of great importance in the design and production of parts in reciprocating machines and components subjected to vibrations. It is always desirable to keep the working stress of material well within the endurance limit.

5.1.11 Strength

The term strength may be defined as the property of a metal by virtue of which it can withstand or support an external force or load without rupture. The strength of a metal is the most important property of metals, and plays a decisive role in designing various structures and components.

Types of strengths

A metal has innumerable types of strengths. The strengths which are important from the current point of view can be broadly grouped into the following two types:

1. Depending upon the values of stress, the strengths of a metal may be elastic or plastic.
2. Depending upon the nature of stress, the strengths of a metal may be tensile, compressive, shear, bending or torsional.

Elastic strength. The elastic strength of a metal is the value of load corresponding to the transition from the elastic to plastic range. The ideal stress values are used to define the elastic strength of a material.

Plastic strength. The plastic strength of a material is the value corresponding to plastic load and rupture. It is also called the ultimate strength. In actual practice, the specimen is subjected to a stress which is always less than the working stress. The ratio of ultimate stress to the working stress of a metal is called the factor of safety. It is also known as the factor of ignorance and greatly varies upon the nature of stresses or loads.

Tensile strength. The tensile strength of a metal is the value of load applied to break it by pulling it outwards into two pieces. The compressive stress is always expressed in $N \ mm^{-2}$ or $MN \ m^{-2}$.

Shear strength. The shear strength of a metal is the value of load applied tangentially to shear it off across the resisting section. The shear stress is expressed in $N \ mm^{-2}$ or $MN \ m^{-2}$.

Bending strength. The bending strength of a metal is the value of load required to break it off by bending it across the resisting action. The bending stress is expressed in $N \ mm^{-2}$ or $MN \ m^{-2}$.

Torsional strength. The torsional strength of a metal is the value of load applied to break it off by twisting across the resisting section. The torsional stress is also expressed in $N \ mm^{-2}$ or $MN \ mm^{-2}$.

5.2 Types of techno-mechanical properties

The technical properties of a metal are those properties which completely define its behavior in shaping, forming and fabrication operation during the manufacturing process. A sound knowledge of these properties is also essential for an engineer to enable him/her to select a suitable metal for the various structures or components of a machine. There are many technological properties of a metal that an engineer should know, but the following are important from our current point of view: malleability, machinability, weldability and castability.

5.2.1 Malleability

The term malleability may be defined as the property of a metal by virtue of which it can be deformed into thin sheets by rolling or hammering without rupture. It depends upon the crystal structure of the metal. Metals with a small grain size are used for very thin sheets, whereas metals with a large grain size are used for thick sheets.

5.2.2 Machinability

The machinability of a metal is defined as the property of the metal which indicates the ease with which it can be cut or removed by cutting tools in various machining operations, such as turning, drilling, boring, milling, etc. The machinability of a metal depends upon the mechanical and physical properties of the metal, its chemical composition, microstructure and cutting conditions.

5.2.3 Weldability

The term weldability can be defined as the property of a metal which indicates the ease with which two similar and dissimilar metals are joined by fusion, with or without the use of a filler metal. Strictly speaking, if a metal has good weldability it can easily be welded in a fabricated structure. The following common metals are listed in descending order of weldability:
1. Iron.
2. Carbon.
3. Steel.
4. Cast iron.
5. Low alloy steel.
6. Stainless steel.

5.2.4 Formability or workability

The term formability or workability can be defined as the property of a metal which indicates the ease with which it can be formed into different shapes and sizes. The various factors affecting the formability of the material are:
- The crystal structure of the metal.
- The grain size of the metal.
- The hot and cold working.
- The alloying element present in the parent metal.

5.2.5 Castability

The term castability can be defined as the property of a metal which indicates the ease with which it is casted into different shapes and sizes from its liquid state. The various factors affecting the castability of a metal are: the solidification rate, gas porosity, segregation and shrinkage.

5.3 Elastic deformation

Deformation in which stress and strain are proportional is called elastic deformation. Elastic deformation is a non-permanent reversible deformation, i.e. when the load/forces are released the body returns to its original configuration (shape and

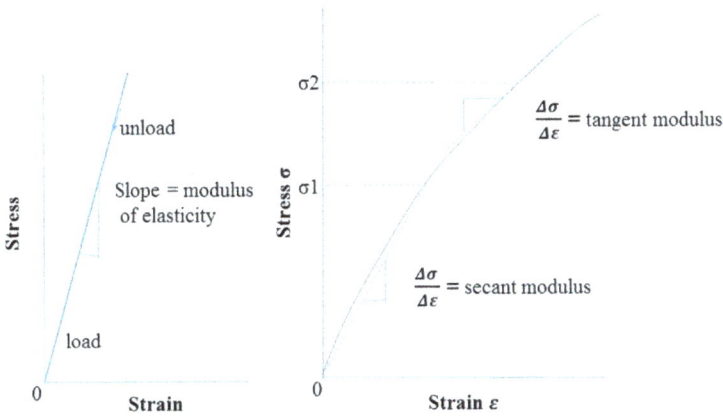

Figure 5.5. Stress–strain diagram in the elastic regions. The methods for the determination of the elastic limits are shown.

size). It can be caused by tension/compression or shear forces. Usually, in metals and ceramics elastic deformation is seen at low strains ($<10^{-3}$). Elasticity can be linear or nonlinear. Metals and ceramics usually show linear elastic behavior. Some materials (e.g. gray cast iron, concrete and many polymers) exhibit nonlinear elasticity.

A stress–strain curve is a graph derived from measuring load (stress, σ) versus extension (strain, ε) for a sample of a material (see figure 5.5). For most metals that are stressed in tension at relatively low levels, stress and strain are proportional to each other, i.e. the slope of this linear segment corresponds to the modulus of elasticity E. This is known as Hooke's law, and the constant of proportionality E (GPa or psi) is the modulus of elasticity, or Young's modulus. This modulus may be thought of as stiffness, or a material's resistance to elastic deformation. The greater the modulus, the stiffer the material, or the smaller the elastic strain that results from the application of a given stress. For materials showing nonlinear elasticity, either a tangent or secant modulus can be used to determine the modulus of elasticity (see figure 5.5).

5.4 Plastic deformation

Plastic deformation in the broadest sense means permanent deformation in the absence of external constraints (forces, displacements).

From an atomic perspective, plastic deformation corresponds to the breaking of bonds with original atom neighbors and then reforming bonds with new neighbors as large numbers of atoms or molecules move relative to one another; upon removal of the stress they do not return to their original positions.

For most metallic materials, elastic deformation persists only to strains of about 0.005. As the material is deformed beyond this point, the stress is no longer proportional to strain (Hooke's law) and permanent/non-recoverable/plastic deformation occurs. A typical stress–strain diagram for a metallic material is given in figure 5.6. This figure shows that the changes occurred at the transition from the elastic to the plastic region.

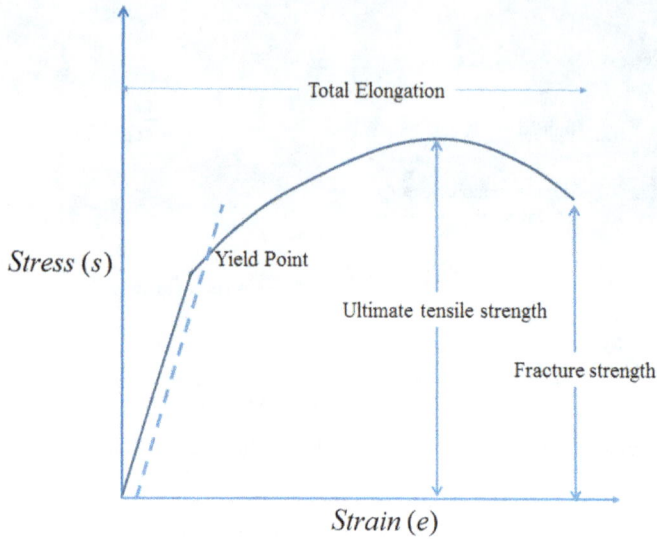

Figure 5.6. Stress–strain diagram for a metallic material. Different regions in the stress–strain diagram are shown.

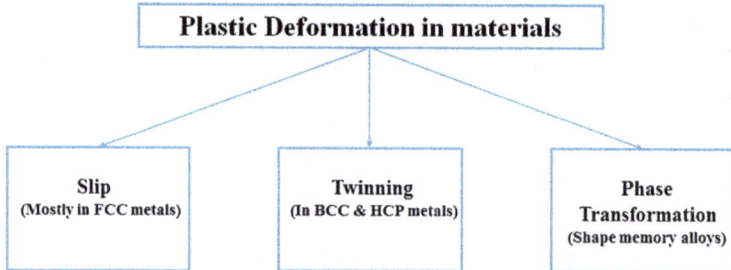

Figure 5.7. Types of plastic deformation in crystalline materials.

The transition from elastic to plastic is a gradual one for most metals. Some curvature occurs at the onset of plastic deformation, which increases more rapidly with rising stress. Plastic deformation in a crystalline solid occurs by means of the various processes described below, among which slip is the most important mechanism. Plastic deformation of crystalline materials takes place by mechanisms (figure 5.7) which are very different from those for amorphous materials (such as glasses). Plastic deformation in amorphous materials occurs by other mechanisms including flow (viscous fluids) and shear banding. The differences between elastic and plastic deformation are listed in table 5.1.

5.5 Slip

Slip is the most important mechanism of plastic deformation. At low temperatures (in particular in bcc metals) twinning may also become important. At the fundamental level plastic deformation (in crystalline materials) by slip involves

Table 5.1. The differences between elastic deformation and plastic deformation.

Elastic deformation	Plastic deformation
1. Appears and disappears with the application and removal of stress.	1. Remains after the removal of stress.
2. The beginning of the process of deformation.	2. Takes place after the elastic deformation has stopped.
3. Takes place over a short range of the stress–strain curve.	3. Takes place over a wide range of the stress–strain curve.
4. The strain reaches its maximum value after the stress has reached its maximum value.	4. The stain occurs simultaneously with the application of stress.

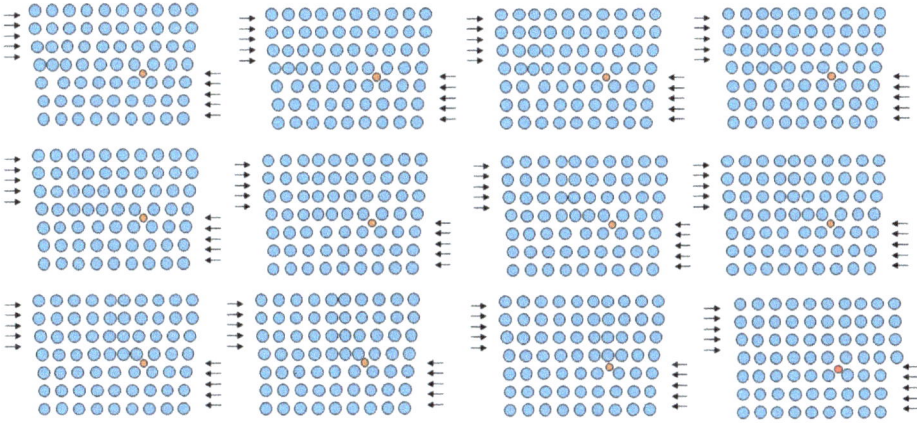

Figure 5.8. Dislocation movement by slip.

the motion of dislocations on the slip plane (creating a step of Burger's vector). Slip is caused by shear stresses (at the level of the slip plane). Hence, a purely hydrostatic state of stress cannot cause slip. A slip system consists of a slip direction lying on a slip plane. Slip is analogous to the mode of locomotion employed by a caterpillar (see figure 5.8).

5.5.1 Slip systems

Dislocations move more easily on specific planes and in specific directions. Ordinarily, there is a preferred plane (slip plane) and specific directions (slip direction) along which dislocations move. An illustration of the movement of dislocations along with the direction of applied stress is given in figure 5.9. The combination of slip plane and slip direction is called the slip system. The slip system depends on the crystal structure of the metal. The slip plane is the plane that has the most dense atomic packing (the greatest PD). The slip direction is most closely packed with atoms (highest LD). In ccp and hcp materials the slip system consists of

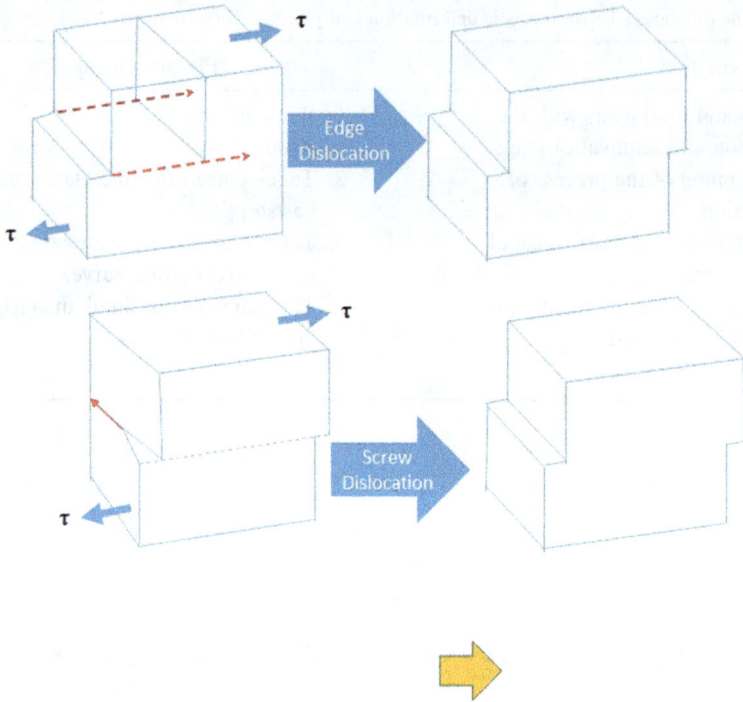

Figure 5.9. Movement of dislocations with respect to the applied stress.

Table 5.2. Slip systems in different crystal systems and compounds.

Crystal	Slip plane(s)	Slip direction	Number of slip systems
fcc	{111}	½ <110>	12
hcp	(0001)	<11$\bar{2}$0>	3
bcc	{110}, {112}, {123}	½ [111]	48
NaCl ionic	{110} {111} not a slip plane	½ <110>	6
C diamond cubic	{111}	½ <110>	12
TiO$_2$ rutile	{101}	<10$\bar{1}$>	
CaF$_2$, UO$_2$, ThO$_2$ fluorite	{001}	<1$\bar{1}$0>	
CsCl	{110}	<001>	
NaCl, LiF, MgO rock salt	{110}	<110>	6
C, Ge, Si diamond cubic	{111}	<110>	12
MgAl$_2$O$_4$ spinel	{111}	<1$\bar{1}$0>	
Al$_2$O$_3$ hexagonal	(0001)	<11$\bar{2}$0>	

a close-packed direction on a close-packed plane. Table 5.2 lists slip systems for different solid crystal systems and compounds. Just the existence of a slip system does not guarantee slip, as a slip is competing against other processes such as twinning and fracture. If the stress to cause slip is very high (i.e. CRSS is very high), then fracture may occur before the slip (as in brittle ceramics). For slip to occur in polycrystalline materials, five independent slip systems are required. Hence, materials which are ductile in the single crystalline form may not be ductile in the polycrystalline form. ccp crystals (Cu, Al, Au) have excellent ductility. At higher temperatures more slip systems may become active and hence polycrystalline materials which are brittle at low temperatures may become ductile at high temperatures.

5.5.2 Slip in a single crystal

If a single crystal of a metal is stressed in tension beyond its elastic limit, it elongates slightly and a step appears on the surface due to the relative displacement of one part of the crystal with respect to the others, and the elongation stops. Further increase in the load causes movement of another parallel plane, resulting in another step. Similarly, a number of small steps are formed on the surface of the single crystal that are parallel to one another and loop around the circumference of the specimen (see figure 5.10). Each step (shear band) results from the movement of a large number of dislocations and their propagation in the slip system.

5.6 Critical resolved shear stress (CRSS)

The slip in crystalline materials results from the action of shear stress on the slip plane. Within the range of shear stresses in engineering applications, the component of the stress normal to the slip plane does not influence slip. Thus the slip process must be considered in terms of the shear stress resolved on the slip plane in the slip direction.

Let, P be equal to the load applied along the axis of the single crystal and A the cross-sectional area of the crystal. As a result of axial load, let the slip take place along the shaded plane as shown in figure 5.11.

Now let α be the angle which the slip direction makes with the tensile axis and β the angle which the slip plane makes with the normal to the tensile axis. We know

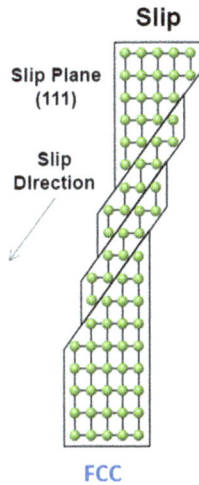

Figure 5.10. Formation of slip bands during deformation by slip in a single crystal.

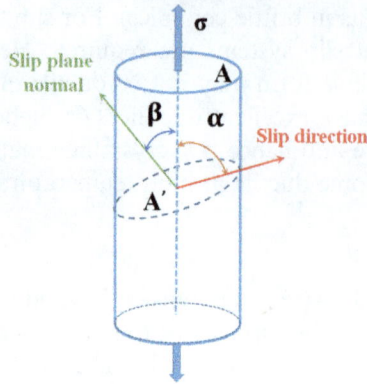

Figure 5.11. Determination of the CRSS from the orientations of the slip direction and the direction applied force.

that the component of the applied load acting in the slip direction is $P \cos\alpha$ and the area of the slip plane is $\dfrac{A}{\cos\beta}$.

The CRSS is

$$\tau = \frac{\text{load}}{\text{area}} = \frac{P\cos\alpha}{A/\cos\beta} = \frac{P}{A}\cos\alpha\cos\beta$$

$$= \sigma\cos\alpha\cos\beta$$

where σ is the applied tensile stress. The stress required to initiate slip in a pure and perfect single crystal is called the CRSS, $\tau_{cr} = \sigma\cos\alpha\cos\beta$. This equation is popularly known as Schmid's law, and the term $\cos\alpha\cos\beta$ as Schmid's factor. The following points are important for the CRSS:

1. If the slip direction is at right angles to the tensile axis (i.e. $\alpha = 90°$), then $\cos\alpha = \cos 90° = 0$. Therefore, $\tau_{cr} = 0$.
2. If the slip plane is parallel to the tensile axis (i.e. $\beta = 90°$) then $\cos\beta = \cos 90° = 0$. Therefore, $\tau_{cr} = 0$.
3. If the slip plane and slip direction are inclined at an angle of 45° to the tensile axis, then

$$\tau_{cr} = \sigma\cos 45°\cos 45° = \sigma \times \frac{1}{\sqrt{2}} \times \frac{1}{\sqrt{2}} = \frac{\sigma}{2}.$$

4. It will be seen that for all combinations of σ and β, the CRSS will always be less than $\sigma/2$ (i.e. half the tensile stress).

The value of the CRSS is a constant for a material at a given temperature. If many slip planes and slip directions of the same type are possible in a crystal, then the active plane is the one on which the CRSS is reached first as the specimen is subjected to increasing applied stress.

5.7 Plastic deformation by twinning

In addition to slip (dislocation movement), plastic deformation can also occur by twinning (see figure 5.12). Twinning results when a portion of the crystal takes up an orientation that is related to the orientation of the rest of the untwined lattice in a definite, symmetrical way. The twinned portion of the crystal is a mirror image of the parent crystal and the plane of symmetry between the two portions is called the twinning plane. Twinning may favorably reorient slip systems to promote dislocation movement. The possible twin planes and directions are given in table 5.3.

Twins are generally of two types: mechanical twins and annealing twins. Mechanical twins are usually observed in bcc and hcp metals and are produced under conditions of rapid loading rate and decreased temperature. Annealing twins are produced as the result of annealing. These twins are usually seen in fcc metals. Annealing twins are usually broader and have straighter sides than mechanical twins (see figure 5.13).

Twinning usually occurs when the slip systems are restricted or when something increases the CRSS so that the twinning stress is lower than the stress for the slip. Therefore, twinning generally occurs at low temperatures or high strain rates in bcc or fcc metals, or in hcp metals. Twinning occurs on specific twinning planes and

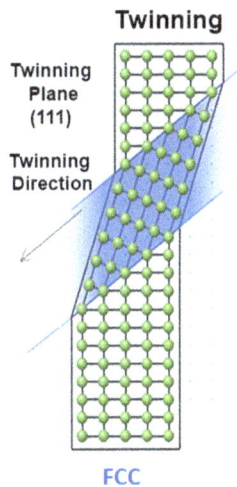

Figure 5.12. A schematic diagram showing how twinning results from an applied shear stress.

Table 5.3. Twin planes and twin directions.

Crystal structure	Typical examples	Twin plane	Twin direction
bcc	α-Fe, Ta	(112)	[111]
hcp	Zn, Cd, Mg, Ti	(10$\bar{1}$2)	[$\bar{1}$011]
fcc	Ag, Au, Cu	(111)	[112]

Figure 5.13. Mechanical twins (nickel base superalloy).

Table 5.4. Difference between slips and twinning.

Slip	Twinning
1. The deformation takes place due to the sliding of atomic planes over others. However, the orientation of the crystals above and below the slip plane is the same after deformation as before.	1. The deformation takes place due to the orientation of one part of the crystal with respect to the other. The twinned portion is the mirror image of the original lattice.
2. The atomic movements are over large distances.	2. The atomic movements are over a fraction of atomic spacing.
3. Requires lower stress for atomic movements.	3. Requires higher stress for atomic movements.
4. Occurs on widely spaced planes.	4. Occurs on an atomic plane involved in the deformation within the twinned region of the crystal.

in twinning directions. The difference between the slip and twinning is given in table 5.4.

5.8 Plastic deformation of polycrystalline materials

Plastic deformation in polycrystalline materials is more complex than that in single crystals due to the presence of grain boundaries, different orientations of neighboring crystals, the presence of several phases, etc. Due to random crystallographic orientations, the slip planes and slip directions vary from one grain to another. As a result, the resolved shear stress τ_{RSS} will vary from one crystal to another and the dislocations will move along the slip systems with favorable orientations (i.e. the

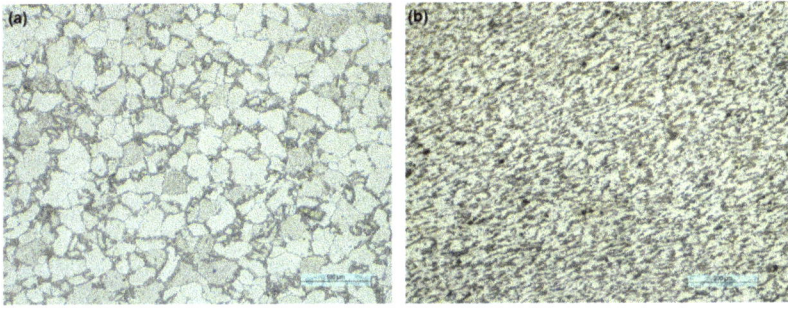

Figure 5.14. Alteration of the grain structure of a polycrystalline metal as a result of plastic deformation: (a) before deformation the grains are equiaxed and (b) after deformation elongated grains are produced

highest resolved shear stress). When a polished polycrystalline specimen of copper is plastically deformed, two slip systems operate for most of the grains (evidenced by two sets of parallel yet intersecting sets of lines). Slip lines are visible, and the variation in grain orientation is indicated by the difference in alignment of the slip lines for several grains. During deformation, mechanical integrity and coherency are maintained along the grain boundaries, i.e. the grain boundaries usually do not come apart or open up. As a consequence, each individual grain is constrained, to some degree, in the shape it may assume by its neighboring grains. The changes brought about in a polycrystalline material due to plastic deformation are shown in figure 5.14.

5.9 Hot working

Hot working is plastic deformation which is carried out under conditions of temperature and strain rate such that substantial recovery processes occur; thus large strains can be achieved with essentially no strain hardening. Hot working is normally performed at a temperature $>0.6T_m$ and high strain rates in the range of 0.5–500 s^{-1}

Advantages of hot working:
- A larger and more rapid deformation can be accomplished since the metal is in the plastic state.
- The porosity of the metal is considerably minimized.
- Concentrated impurities, if any, in the metal are disintegrated and distributed throughout the metal.
- The grain structure of the metal is refined and the physical properties are improved.
- No residual stresses remain in the material

Disadvantages of hot working:
- A poor surface finish and loss of metal due to rapid oxidation or scale formation on the metal surface.
- Close tolerances cannot be maintained.
- Involves excessive expenditure due to the high cost of tooling. This, however, is compensated by the high production rate and better quality of products.

5.10 Warm working

Warm working is the plastic deformation of a metal at a temperature below the temperature range of recrystallization and above room temperature, i.e. intermediate to hot and cold working.

Advantages compared to cold working:
- Combines the advantages of both hot and cold working into one operation.
- Fewer annealing operations (due to less strain hardening).
- Smaller loads on tooling and equipment.
- Greater metal ductility.

Advantages compared to hot working:
- Improved dimensional control.
- Lower energy costs.
- Better precision of components.
- Smaller scaling and decarburization on parts.
- A better surface finish.
- Smaller thermal shock on tooling.
- Smaller thermal fatigue to tooling and thus greater life of tooling.

5.11 Cold working

Plastic deformation carried out at a temperature region and over a time interval, such that the strain hardening is not relieved is called cold work. It is normally performed at room temperature, but in general at $<0.3T_m$, where recovery is limited and recrystallization does not occur. The detailed differences between hot working and cold working are given in table 5.5.

Advantages of cold working:
- Due to work hardening the strength and hardness of the metal are increased.
- It is an ideal method to increase the hardness of those metals which do not respond to heat treatment.
- Better dimensional control is possible because the reduction in size is not much.
- Provides a fine grain size and good surface finish (no oxidation takes place).
- Handling is easier because of the low operating temperature.
- Directional properties can be imparted.

Disadvantages of cold working:
- Only ductile metals can be shaped through cold working.
- Over-working of the metal results in brittleness and it has to be annealed to remove the same.
- Subsequent heat treatment is usually needed to remove the residual stresses set up during cold working.
- Higher forces and heavier and more powerful equipment are required for deformation.

Table 5.5. Differences between cold working and hot working. Theoretically, hot working = cold working + annealing.

Cold working	Hot working
1. Working of metals and alloys below their recrystallization temperature.	1. Working of metals and alloys above their recrystallization temperature.
2. Strain hardening occurs during cold working. Due to this the tensile strength and hardness increase while the impact strength and ductility decrease.	2. Strain hardening is removed by recrystallization which occurs at high temperature and hence no property change is observed during hot working.
3. Microstructure shows distorted grains.	3. Microstructure shows equiaxed and usually refined grains.
4. Defect density increases, i.e. vacancies, dislocations, etc, increase and hence the density of the material slightly decrease.	4. Almost no change in defect density and hence no change in the density of the material. If blow holes and other such defects are originally present, they are largely eliminated and under this situation the density may arise.
5. Working cannot be carried out indefinitely without cracking of the material due to strain hardening.	5. Working can be done to any extent without cracking due to elimination of strain hardening at higher temperatures.
6. The energy required for plastic deformation is greater.	6. The energy required for plastic deformation is lower because at higher temperatures metals become soft and ductile.
7. More stress is required for deformation.	7. Less stress is required for deformation.
8. Retains chemical heterogeneity if present; this is usually present in cast components.	8. Reduces the chemical heterogeneity of cast components due to faster diffusion at elevated temperatures.
9. No oxidation of metals occurs during working and hence pickling is not required.	9. Heavy oxidation occurs during working and pickling is required to remove the oxides.
10. Embrittlement does not occur due to less diffusion, and no reaction of oxygen at lower temperatures.	10. Reactive metals become severely embrittled by oxygen and hence must be protected from the action of oxygen by using an inert atmosphere.
11. Surface decarburization in steels does not occur.	11. Surface decarburization in steels is likely to occur at higher temperatures unless the steel is protected by a proper atmosphere.
12. The surface finish is good.	12. The surface finish is not as good due to oxidation at higher temperatures.
13. It is easy to control the dimensions within the tolerance limit.	13. It is difficult to control the dimensions because of the contractions occurring during cooling.
14. Handling of materials is easy.	14. Handling materials is difficult.
15. Ordinary steels can be used for shaping and hence the cost of cold working plants is lower.	15. Alloy steels are necessary for shaping and hence the cost of the hot working plants is high.

5.12 Recovery, recrystallization and grain growth

The plastic deformation in the temperature range $(0.3–0.5)T_m$ is called cold working. Point defects and dislocations have associated strain energy and 1–10% of the energy expended in plastic deformation is stored in the form of strain energy (in these defects), thus the *material becomes a battery of energy*. The cold worked material is in a microstructurally metastable state. Depending on the severity of the cold work the dislocation density can increase by 4–6 orders of magnitude. The material becomes stronger, but less ductile. The cold worked material is stronger (harder), but is brittle. Heating the material (typically below $0.5T_m$) and holding for a sufficient time is a heat treatment process called annealing. Depending on the temperature of annealing, processes such as recovery (at lower temperatures) or recrystallization (at higher temperatures) may take place. During these processes the material tends to go from a microstructurally metastable state to a lower energy state (towards a stable state). Further annealing of the recrystallized material can lead to grain growth:

$$\text{annealed material} \xrightarrow[\rho_{\text{dislocation}} \sim (10^{12}-10^{14})]{\text{coldwork}} \text{stronger material}$$

$$\rho_{\text{dislocation}} \sim (10^6-10^9) \qquad \rho_{\text{dislocation}} \sim (10^{12}-10^{14}).$$

5.12.1 Recovery

The term 'recover' may be defined as the process of removing the internal stresses in a metal by heating it to a relatively low temperature, which is usually below the melting point. It has been observed that the recovery process does not affect the grain structure, but it removes the internal stresses only. Moreover, the recovery processes do not affect the hardness and strength, but increase the ductility of the metal. As a result of cold working the dislocations pile up at the grain boundaries. During the processes of recovery, these dislocation start reducing and rearranging themselves. They do so through a mechanism known as polygonization (see figure 5.15). In this mechanism, the dislocations climb out of their slip planes and rearrange themselves in a lower energy configuration.

5.12.2 Recrystallization

The term 'recrystallization' may be defined as the process of forming strain free new grains in a metal by heating it to a temperature known as the recrystallization temperature. It may be noted that the recrystallization temperature is, usually, the

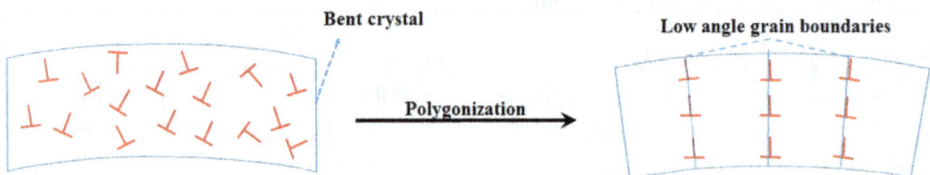

Figure 5.15. Rearrangement of dislocations into low-energy orientations by polygonization.

temperature at which about 50% of cold worked metal recrystallizes in one hour. It has been observed that the formation of new grains in recrystallization takes place through the following three processes:

- Nucleation.
- Primary grain growth.
- Secondary grain growth.

During nucleation, small strain free nuclei are formed at a grain boundary where the deformation is very high (see figure 5.16). These nuclei grow into strain free grains during primary grain growth. These grains meet each other and replace the old grains by the new ones. During the secondary grain growth, these new grains grow at the expense of others. At the end of this stage the grains are, usually, of very small size but of the same shape. The factors which affect the recrystallization processes are the time of heating, the temperature prior to deformation, impurities, alloy additions, etc.

It has been observed that an increase in the time of heating or annealing reduces the recrystallization temperature. A certain amount of deformation is required before the crystallization may take place. The addition of impurities or alloys increases the recrystallization temperature. It should be noted that the recrystallization process does not produce new structures, but produces strain-free new grains and achieve complete elimination of internal stresses. This results in a sharp decrease in the hardness and strength as well as increasing ductility.

5.12.3 Grain growth

The term 'grain growth' may be defined as the processes of forming larger strain free grains in the metal through heating, at a temperature above that of recrystallization. It should be noted that recrystallization produces strain free new grains. These grains are of smaller size, but of the same shapes. When the temperature is

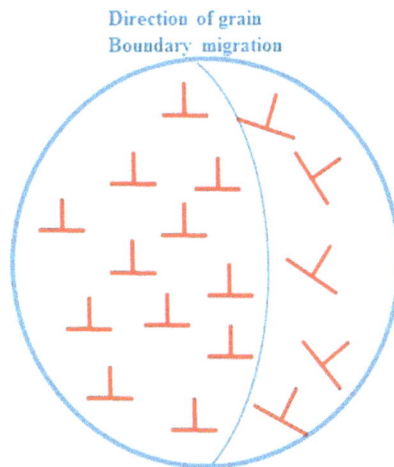

Figure 5.16. Growth of grains in the direction of higher dislocation density.

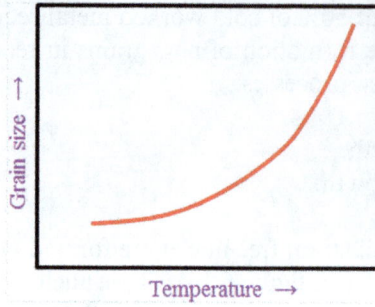

Figure 5.17. Effect of temperature on recrystallized grain size.

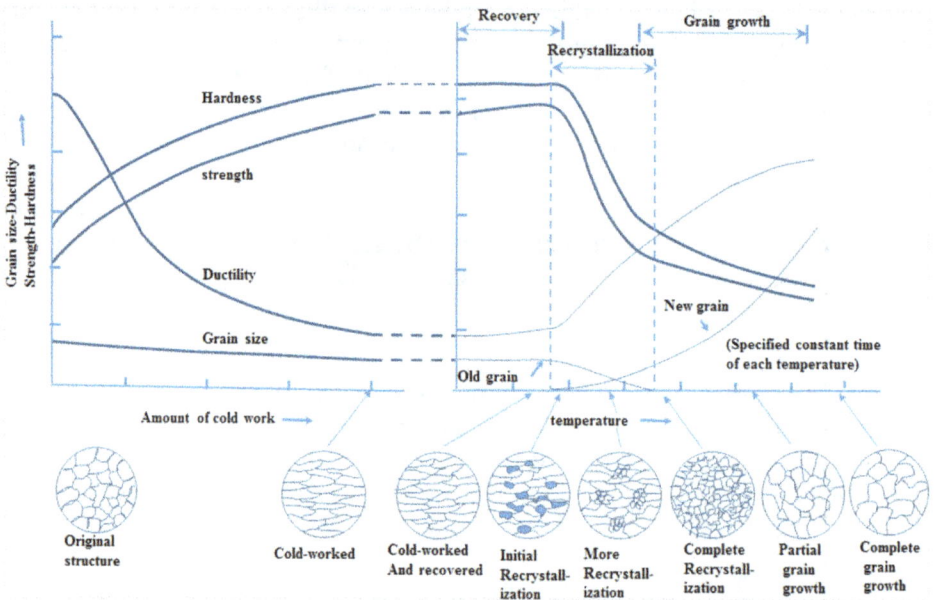

Figure 5.18. Variation of mechanical properties at different stages.

increased above that of recrystallization, these grains grow in size. The growth of grains takes place even during recrystallization, but the growth rate is slow and only becomes rapid with an increase of temperature (see figure 5.17). Grain growth takes place due to the combination of individual grains, thereby reducing the grain boundary area. As a result of this, the total energy decreases and the grains become stable. The factors which affect the growth rate are the time of heating, temperature, degree of cold work and addition of impurities. It has been observed that the grain growth results in the decrease of hardness and strength as well as an increase in ductility.

The mechanical properties of a metal undergo variation with annealing treatment at different stages. This is illustrated in figure 5.18.

Further reading

ASM International 1990 *Irons and Steels and High Performance Alloys: Specialty Steels and Heat Resistant Alloys* (*Metals Handbook* vol 1) 10th edn (Materials Park, OH: ASM International) pp 755–1003

Avner S H 1997 *Introduction to Physical Metallurgy* (New York: McGraw-Hill)

Barett C S 1952 *Structure of Metals* 2nd edn (New York: McGraw-Hill)

Boas W 1947 *An Introduction to the Physics of Metals and Alloys* (New York: Wiley)

Brick R M and Phillips A 1949 *Structure and Properties of Alloys* 2nd edn (New York: McGraw-Hill)

Cottrel A H 1956 *Dislocations and Plastic Flow in Crystals* (Fair Lawn, NJ: Oxford University Press)

Guy A G 1959 *Elements of Physical Metallurgy* 2nd edn (Reading, MA: Addison-Wesley)

Hirth J P and Lothe J *Theory of Dislocations* (New York: McGraw-Hill)

Honeycombe R W K 1984 *Plastic Deformation of Metals* (London: Edward Arnold)

Kelly A, Groves G W and Kidd P 2000 *Crystallography and Crystal Defects* (Chichester: Wiley)

Khurmi R S and Sedha R S 2000 *Material Science* (New Delhi: S Chand and Company Limited)

Kingery W D, Bowen H K and Uhlmann D R 1976 *Introduction to Ceramics* (New York: Wiley) ch 14

Raghavan V 2004 *Materials Science and Engineering* 5th edn (Englewood Cliffs, NJ: Prentice-Hall)

Reed-Hill R E 1964 *Physical Metallurgy Principles* (New York: Van Nostrand Reinhold)

Subramaniam A and Balani K (IITK) *Materials Science and Engineering* (e-book) MHRD, India

Chapter 6

The theory of alloys

6.1 The concept of alloy formation

An alloy is a substance that has metallic properties and is composed of two or more chemical elements, of which at least one is a metal. An alloy system contains all the alloys that can be formed by several elements combined in all possible proportions. If the system is made up of two elements, it is called a binary alloy system; three elements, a ternary alloy system; etc. Taking only 45 of the most common metals, any combination of two gives 990 binary systems. Combinations of three give over 14 000 ternary systems. However, in each system, a large number of different alloys are possible. If the composition varies by 1%, each binary system will yield 100 different alloys. Since commercial alloys often contain many elements, it is apparent that the number of possible alloys is almost infinite. Alloys may be homogeneous (uniform) or mixtures. If the alloy is homogeneous it will consist of a single phase, and if it is a mixture it will be a combination of several phases. The uniformity of an alloy phase is not determined on an atomic scale, such as the composition of each unit lattice cell, but rather a much larger scale. Figure 6.1 can be considered as a possible classification of alloy structures.

6.2 Phase

A phase is anything which is homogeneous and physically distinct. Any structure which is visible as physically distinct microscopically may be considered as a phase. For most pure elements the term *phase* is synonymous with state. Some metals are allotropic in the solid state and will have different solid phases. When the metal undergoes a change in crystal structure, it undergoes a phase change since each type of crystal structure is physically distinct. In a pure material, when other elements are added intentionally they are called alloying elements. Alloying elements are added to improve certain properties of the pure element. The alloying element can be accommodated in one of the three possibilities shown in figure 6.2.

doi:10.1088/978-1-6817-4473-5ch6   6-1

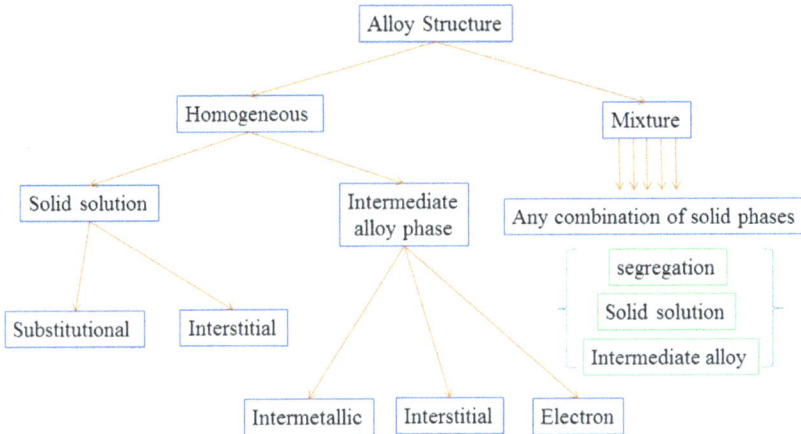

Figure 6.1. Broad classification of alloy structures.

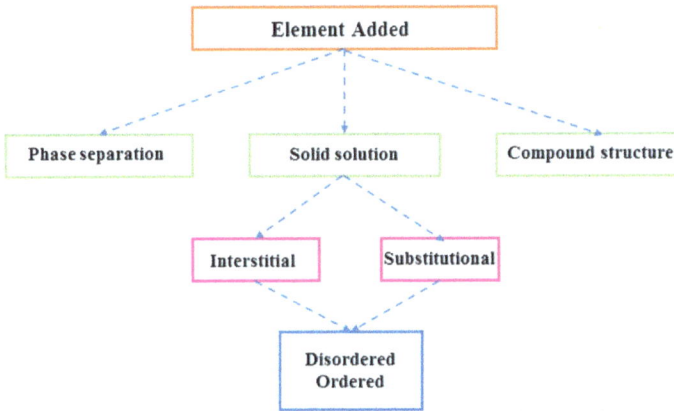

Figure 6.2. Possible accommodation of alloying elements in an alloy system.

6.3 Segregation/phase separation

If the added element does not dissolve in the parent/matrix phase it may form a separate phase. In a polycrystal it may go to the grain boundary and may segregate to other defects such as dislocation cores, etc. The solubility in the case of a substitutional solid solution is given by Hume-Rothery rules (considered below).

6.4 Solid solutions

Any solution is composed of two parts: a solute and a solvent. The solute is the minor part of the solution or the material which is dissolved, while the solvent constituents the major portion of the solution. The most common solutions involve water as the solvent, such as sugar or salt dissolved in water. There are three possible conditions for a solution: unsaturated, saturated and supersaturated. If the solvent is dissolving less of the solute than it could dissolve at a given temperature and pressure, it is said to be unsaturated. If it is dissolving the limiting amount of solute, it is saturated. If it is dissolving more of the solute than it should under equilibrium conditions, the solution is supersaturated. The supersaturated condition is an unstable one, and given enough time or a little energy, the solution tends to become stable or saturated by rejecting or precipitating the excess solute. A solid solution is simply a solution in the solid state and consists of two kinds of atoms combined in one type of space lattice. There are two types of solid solutions, substitutional and interstitial.

6.4.1 Substitutional solid solutions

In this type of solution, the atoms of the solute substitute for atoms of the solvent in the lattice structure of the solvent. For example, silver atoms may substitute for gold atoms without losing the fcc structure of gold, and gold atoms may substitute for silver atoms in the fcc lattice structure of silver. Several factors are now known, largely through the work of Hume-Rothery, that control the range of solubility in alloy systems.

6.4.2 The Hume-Rothery rules

Crystal-structure factor. The complete solid solubility of two elements is never attained unless the elements have the same type of crystal lattice structure.

Relative-size factor. The size factor is favorable for solid solution formation when the difference in atomic radii is less than about 15%.

Valency rule. A metal will dissolve a metal of higher valency to a greater extent than one of lower valency. The solute and solvent atoms should typically have the same valance in order to achieve maximum solubility.

Electronegativity rule. An electronegativity difference close to zero gives the maximum solubility. The more electropositive one element and the more electro-negative the other, the greater is the likelihood that they will form an inter-metallic compound instead of a substitutional solid solution. The solute and the solvent should lie relatively close in the electrochemical series. The different deciding factors for the solubility of a single element are listed in table 6.1.

Table 6.1. A comparison of different factors that favor substitutional solid solutions.

System		Crystal structure	Radius of atoms (Å)	Valency	Electronegativity
Ag–Au	Ag	fcc	1.44	1	1.9
	Au	fcc	1.44	1	2.4
Cu–Ni	Cu	fcc	1.28	1	1.9
	Ni	fcc	1.25	2	1.8
Ge–Si	Ge	dc	1.22	4	1.8
	Si	dc	1.18	4	1.8

6.4.3 Interstitial solid solutions

The second species added goes into the voids of the parent lattice, e.g. octahedral and tetrahedral voids, in ccp, hcp and bcc crystals (e.g. of solvents Fe, Mo, Cr, etc), e.g. B ($r = 0.97$ Å) C ($r = 0.77$ Å), N ($r = 0.71$ Å), O ($r = 0.66$ Å) and H ($r = 0.46$ Å). The solubility of interstitial atoms is greater in transition elements (Fe, Ti, V, Zr, Ni, W, U, Mn, Cr). Due to its electronic structure (incomplete inner shell) C is particularly insoluble in most non-transition elements

6.5 Compound/intermediate structure

Chemical compounds are a combination of positive and negative valence elements. Intermetallic compounds can be very different from normal chemical compounds (e.g. H_2O). Most compounds are like pure metals, so the cooling curve for a compound is similar to that for a pure metal. A broad classification of chemical compounds is given in figure 6.3.

6.5.1 Valency compounds

Compounds have a different crystal lattice compared to their components. Most chemical compounds have complex crystal structures. Each component has a specific location in the lattice. The composition can be specified by a simple formula:

$$A_n B_m: Mg_2Sn, Mg_2Pb, Ni_3(Al, Ti),$$

where m and n are small whole numbers. Valency compounds also have different properties than their components, such as a constant melting point and dissociation temperature. They are accompanied by s substantial thermal effect and are typically formed by elements with very different electronic and crystal structures. The bonding in intermetallic compounds is usually metallic. The bonding between a metal and a non-metal can also be metallic. A large number of intermetallic compounds do not obey valency rules or have a constant composition (thus distinguishing them from the usual chemical compounds).

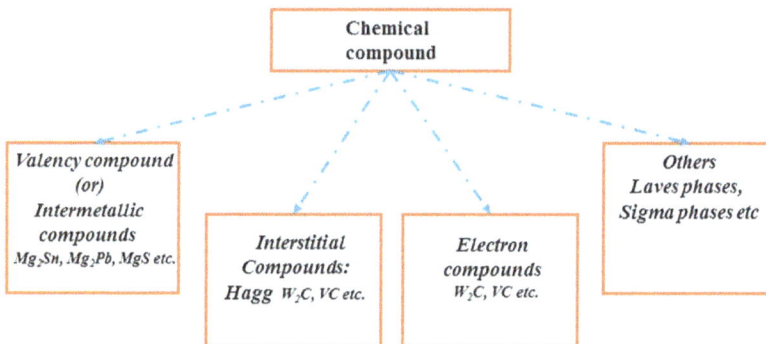

Figure 6.3. Classification of various chemical compounds.

6.5.2 Interstitial compounds

Transition metals form compounds with elements with a small atomic size (H, C, N, B):

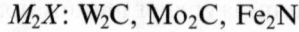

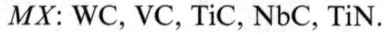

$$M_4X: Fe_4N, Mn_4N$$

$$M_2X: W_2C, Mo_2C, Fe_2N$$

$$MX: WC, VC, TiC, NbC, TiN.$$

The crystal structure depends on the R_x/R_M ratio. If $R_x/R_M < 0.59$ one obtains simple crystal lattices (cubic, hexagonal). The non-metal occupies specific interstitial sites in the cubic or hexagonal crystal. If $R_x/R_M > 0.59$ one obtains complex crystal structures (e.g. Fe_3C). Apart from the size factor, the valency of the interstitial atom also seems to play some role. Typically, interstitial phases have a variable composition; the chemical formula indicates the maximum amount of non-metal in the structure.

6.5.3 Electron compounds

The compounds have a definite ratio of valence electrons to atoms and are therefore called electron compounds. They are usually formed in:
- Monovalent metals (Cu, Ag, Au, Li, Na).
- Transition metals (Fe, Co, Mn).
- Other metals with a valency value between 2 and 5.

Alloys of Cu, Ag and Au frequently form electron compounds. These compounds have specific ratios for the number of valence electrons/atoms:
- 3:2 for bcc (e.g. AgCd, AgZn, Cu_3Al, FeAl, etc).
- 21:13 for ccp (e.g. Ag_5Cd_8, Cu_9Al_4, Fe_5Zn_{21}, Ni_5Zn_{21}, etc).
- 7:4 for hcp (e.g. $AuZn_3$, $AgCd_3$, Cu_3Si, Ag_3Sn).

These compounds can form over a wide range of compositions. For example, in the compound AgZn, the atom of silver has one valence electron while that of zinc has two valence electrons, so the two atoms of the compound will have three valence electrons, or an electron-to-atom ratio of 3:2. Consider that in the compound Cu_9Al_4 each atom of copper has one valence electron and each atom of aluminum has three valence electrons, so that the 13 atoms that make up the compound have 21 valence electrons, or an electron-to-atom ratio of 21:13.

6.6 Other phases

Laves phases. These have a general formula of AB_2, for example $MgCu_2$ (cubic), $MgZn_2$ (hexagonal) and $MgNi_2$ (hexagonal).

Sigma phases. These have a very complex crystal structure and are very brittle. They can act as a source of embrittlement in some alloys such as steels.

Further reading

Avner S H 1997 *Introduction to Physical Metallurgy* (New York: McGraw-Hill)

Callister W D 2007 *Callister's Materials Science and Engineering* (Indian Adaptation adapted by R Balasubramaniam) (New Delhi: Wiley)

Porter D A, Eastererling K E and Sherif M Y 2009 *Phase Transformations in Metals and Alloys* (Boca Raton, FL: CRC Press Taylor and Francis Group)

Raghavan V 2004 *Materials Science and Engineering* 5th edn (Englewood Cliffs, NJ: Prentice-Hall)

Subramaniam A and Balani K (IITK) *Materials Science and Engineering* (e-book) MHRD, India

Chapter 7

Phase diagrams

Phase diagrams are an important tool in the armory of a materials scientist. In the simplest sense a phase diagram demarcates the regions of existence of various phases; phase diagrams are maps. Phase diagrams are also referred to as 'equilibrium diagrams' or 'constitutional diagrams'. This usage requires special attention; although the term used is 'equilibrium', in practical terms the equilibrium is *not a global equilibrium* but a *microstructure-level equilibrium*. Broadly, two kinds of phase diagrams can be differentiated: those involving time and those which do not involve time. In this chapter we shall deal with the phase diagrams that do not involve time. This type can be further sub-classified into those phases with composition as a variable (e.g. T versus % composition) and those without the composition as a variable (e.g. P versus T). Time–temperature transformation (TTT) diagrams and continuous-cooling transformation (CCT) diagrams involve time and will be considered in sections 8.3 and 8.4.

7.1 Basic definitions

Components of a system. These are independent chemical species which comprise the system. These could be elements, ions and/or compounds. Examples:
- Au–Cu system: components Au, Cu (elements).
- Ice-water system: component H_2O (compound).
- Al_2O_3–Cr_2O_3 system: components Al_2O_3, Cr_2O_3 (compounds).

Phase. A physically homogeneous and distinct portion of a material system (e.g. gas, crystal, amorphous…).
- Gases: the gaseous state is always a single phase, it is mixed at the atomic or molecular level.
- Liquids: a liquid solution is a single phase, e.g. NaCl in H_2O, and liquid mixtures consist of two or more phases, e.g. oil in water (no mixing at the atomic level).

doi:10.1088/978-1-6817-4473-5ch7

- Solids: in general, due to the multiple compositions and crystal structures many phases are possible. For the same composition different crystal structures represent different phases, e.g. Fe (bcc) and Fe (fcc) are different phases.

Phase transformation. Phase transformation is the change of one phase into another. For example:
- water → ice.
- α-Fe (bcc) → γ-Fe (fcc).

Grain. The single crystalline part of polycrystalline metal separated by similar entities by a grain boundary.

Solute. The component of either a liquid or solid solution that is present to a lesser or minor extent; the component that is dissolved in the solvent.

Solvent. The component of either a liquid or solid solution that is present to a greater or major extent; the component that dissolves the solute.

System. This term has two meanings. First, 'system' may refer to a specific body of material or object. Second, it may relate to the series of possible alloys consisting of the same components, but without regard to alloy composition.

Solubility limit. For many alloy systems and at some specific temperature, there is a maximum concentration of solute atoms that may dissolve in the solvent to form a solid solution; this is called the solubility limit.

Microstructure. Phases + defects + residual stress and their distributions. Structures requiring magnifications in the region of 100–1000 times, or the distribution of phases and defects in a material.

Phase diagram. A map that gives the relationships between phases in equilibrium in a system as a function of T, P and composition. A map demarcating regions of stability of various phases.

Variables/axes of phase diagrams. The axes can be:
- Thermodynamic (T, P, V).
- Kinetic (t) or composition variables (C, %X).

In single-component systems (unary systems) the usual variables are T and P. In the phase diagrams used in materials science, the usual variables are T and %X. In the study of phase transformation kinetics TTT diagrams or CCT diagrams are also used where the axes are T and t.

System components. Phase diagrams and the systems they describe are often classified and named for the number (in Latin) of components in the system, see table 7.1.

Coordinates of phase diagrams. Phase diagrams are usually plotted with temperature, in degrees centigrade or Fahrenheit, as the ordinate and the alloy composition in weight percentage as the abscissa.

Table 7.1. The number of components present and the corresponding names given to the phase diagram.

Number of components	Name of system or diagram
one	unary
two	binary
three	ternary
four	quaternary
five	quinary
six	sexinary
seven	septenary
eight	octanary
nine	nonary
ten	decinary

The weight % of component A: $W_A = \dfrac{\text{weight of component A}}{\Sigma \text{ weight of all components}} \times 100.$

The atom (or mol)% of component A:

$$X_A = \frac{\text{number of atoms (or mols) of component A}}{\Sigma \text{ number of atoms (or mols) of all components}} \times 100.$$

The conversion from weight percentage to atomic percentage may be made by the following formulas:

$$\text{atomic \% of A} = \frac{X}{X + Y\left(\dfrac{M}{N}\right)} \times 100$$

$$\text{atomic \% of B} = \frac{Y\left(\dfrac{M}{N}\right)}{X + Y\left(\dfrac{M}{N}\right)} \times 100$$

where M is the atomic weight of metal A, N is the atomic weight of metal B, X is the weight percentage of metal A and Y is the weight percentage of metal B.

7.2 Gibbs phase rule

The phase rule connects the degrees of freedom, the number of components in a system and the number of phases present in a system via a simple equation. To understand the phase rule, one must understand the variables in the system along with the degrees of freedom.

Degrees of freedom. The degree of freedom, F, are those externally controllable conditions of temperature, pressure and composition, which are independently variable and which must be specified in order to completely define the equilibrium state of the system.

For a system in equilibrium

$$F = C - P + 2.$$

Where F is the degrees of freedom, C is the number of components and P is the number of phases. The degrees of freedom cannot be less than zero, so we have an upper limit to the number of phases that can exist in equilibrium for a given system.

7.3 Unary phase diagram

Let us start with the simplest system possible: the unary system wherein there is just one component.

Although there are many possibilities, even in the unary phase diagram (in terms of the axis and phases), we shall only consider a T–P unary phase diagram. Let us consider the unary phase diagram of water (H_2O) (see figure 7.1). The Gibbs phase rule here is: $F = C - P + 2$ (2 is for T and P; no composition variables here). Along the two-phase co-existence (at B and C) lines the degree of freedom (F) is 1, i.e. we can chose either T or P and the other will be fixed automatically. The three-phase co-existence points (at A) are invariant points with $F = 0$ (the invariant point implies they are fixed for a given system). The single-phase region at points D, T and P can both be varied while still being in the single-phase region with $F = 2$.

Figure 7.2 represents the phase diagram for pure iron. The triple point temperature and pressure are 490 °C and 110 kbars, respectively. α, γ and ε refer to ferrite, austenite and ε-Fe, respectively. δ is simply the higher temperature designation of α.

7.4 Binary phase diagram

Binary implies that there are two components. Pressure changes often have little effect on the equilibrium of solid phases (unless of course we apply 'huge' pressures). Hence,

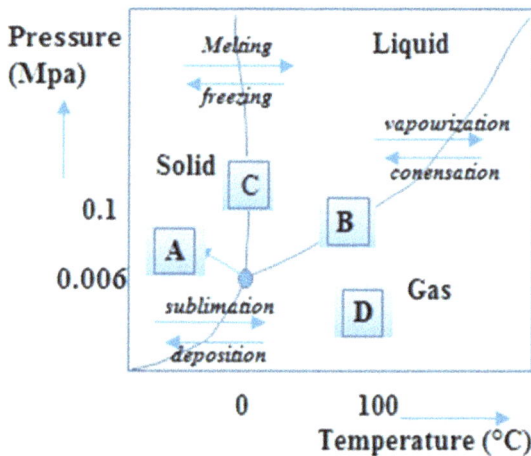

Figure 7.1. Unary phase diagram of water.

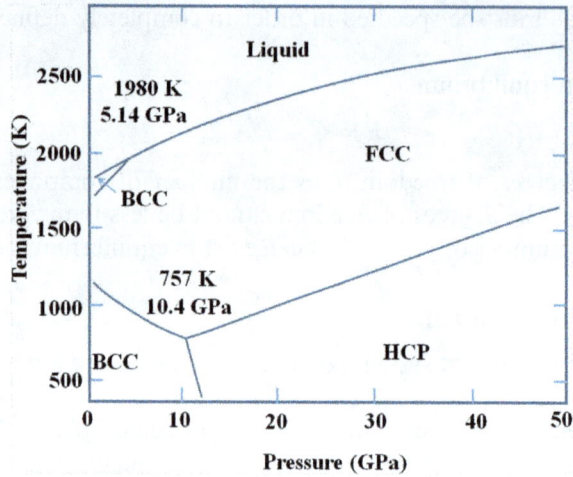

Figure 7.2. Unary phase diagram of iron. Reproduced from Bhadeshia and Honeycombe (2006) with permission of Elsevier.

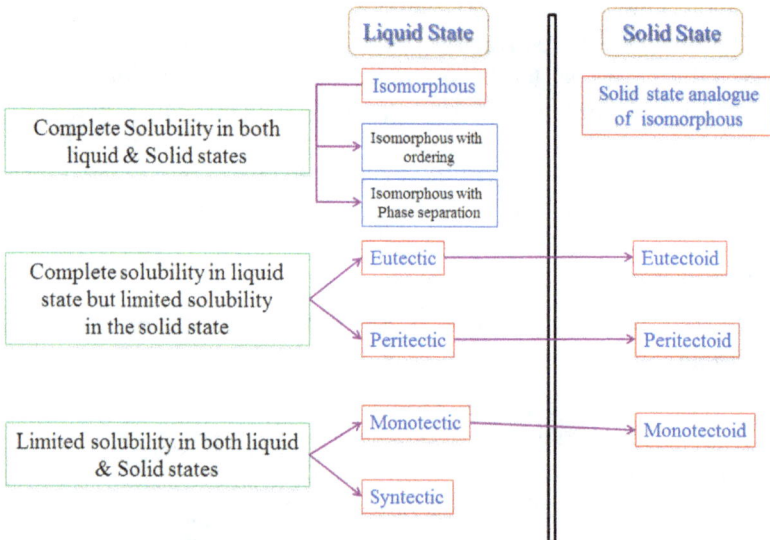

Figure 7.3. Types of phase diagrams based on the solubility of the solute and solvent atoms in the solid phase and liquid phase. Reproduced with permission of Professor Subramaniam, Indian Institute of Technology http://home.iitk.ac.in/~anandh/E-book/.

binary phase diagrams are usually drawn at a pressure of one atmosphere. The Gibbs phase rule is reduced to $F = C - P + 1$ (1 is for T) and the variables are reduced to T and composition (these are the usual variables in materials phase diagrams).

In figure 7.3 we consider the possible binary phase diagrams. These have been classified based on:

- Complete solubility in both the liquid and solid states.

- Complete solubility in both the liquid state, but limited solubility in the solid state.
- Limited solubility in both the liquid and solid states.

7.4.1 Isomorphous phase diagram

Isomorphous phase diagrams form when there is complete solid and liquid solubility (see figure 7.4). Complete solid solubility implies that the crystal structures of the two components have to be same and the Hume-Rothery rules need to be followed. Examples of systems that form isomorphous systems are: Cu–Ni, Ag–Au, Ge–Si and Al_2O_3–Cr_2O_3. Both the liquid and solid contain the components A and B. In binary phase diagrams between two single-phase regions there will be a two-phase region; in the isomorphous diagram between the liquid and solid state, there is the (liquid + solid) state. The liquid + solid state is *not* a semi-solid state, it is a solid of fixed composition and structure, in equilibrium with a liquid of fixed composition. In some systems (e.g. the Au–Ni system) there might be phase separation in the solid state (i.e. the complete solid solubility criterion may not be followed), these will be considered as variations of the isomorphous system below.

Chemical composition of phases (tie line rule)
To determine the actual chemical composition of the phases of an alloy in equilibrium at any specified temperature in a two-phase region, draw a horizontal temperature line, called a tie line, to the boundaries of the field. These points of intersection are dropped to the base line and the composition is read directly (see figure 7.5).

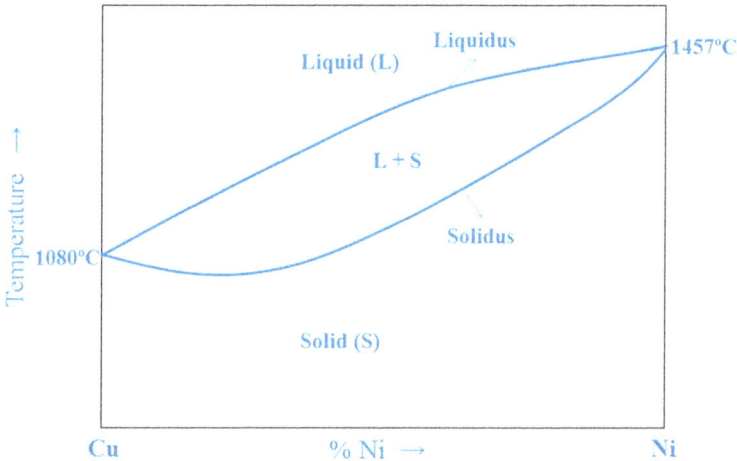

Figure 7.4. Isomorphous phase diagram of the Cu–Ni alloy system.

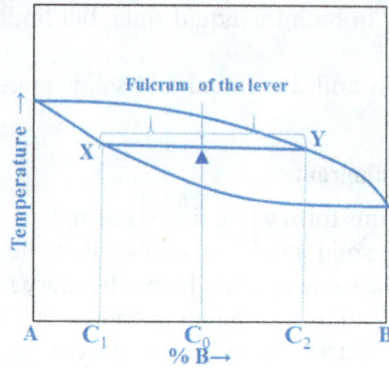

Figure 7.5. Tie line rule for the determination of the chemical composition of phases.

Relative amounts of each phase (lever rule)

To determine the relative amounts of the two phases in equilibrium at any specified temperature in a two-phase region, draw a vertical line representing the alloy and a horizontal temperature line to the boundaries of the field. The vertical line will divide the horizontal line into two parts whose lengths are inversely proportional to the amount of the phases present. This is also known as the *lever rule*. The point where the vertical line intersects the horizontal line may be considered as the *fulcrum* of a lever system. The relative lengths of the lever arms multiplied by the amounts of the phases present must balance.

We draw a horizontal line (called the tie line) at the temperature of interest (say T_0). Let the tie line be XY. A solid (crystal) of composition C_1 coexists with a liquid of composition C_2. Note that tie lines can be drawn only in the two-phase coexistence regions (fields), although they may be extended to mark the temperature. To find the fractions of solid and liquid we use the lever rule. The portion of the horizontal line in the two-phase region is akin to a 'lever' with the fulcrum at the nominal composition (C_0). The opposite arms of the lever are proportional to the fraction of the solid and liquid phase present (this is the lever rule)

$$f_{\text{liquid}} = \frac{C_0 - C_1}{C_2 - C_1} \quad f_{\text{solid}} = \frac{C_2 - C_0}{C_2 - C_1}$$

Variations of the isomorphous system

Elevation of the melting point means that the solid state is 'more stable' and the ordering reaction is seen at low T. This indicates that A–B bonds are stronger than A–A and B–B bonds and the solid favors an ordered solid formation.

Depression of the melting point indicates that the liquid state (disordered) is more stable and phase separation is seen at low T (phase separation can be thought of as the opposite of ordering). The A–A and B–B bonds are stronger than A–B bonds and the liquid favors phase separation in the solid. These kinds of variations are shown in figure 7.6.

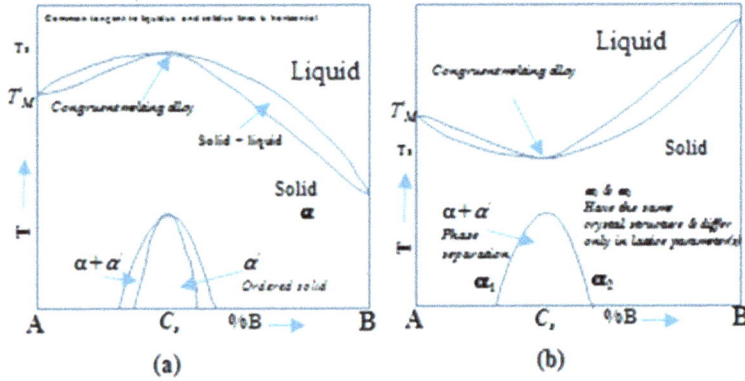

Figure 7.6. Variation in the isomorphous phase diagram.

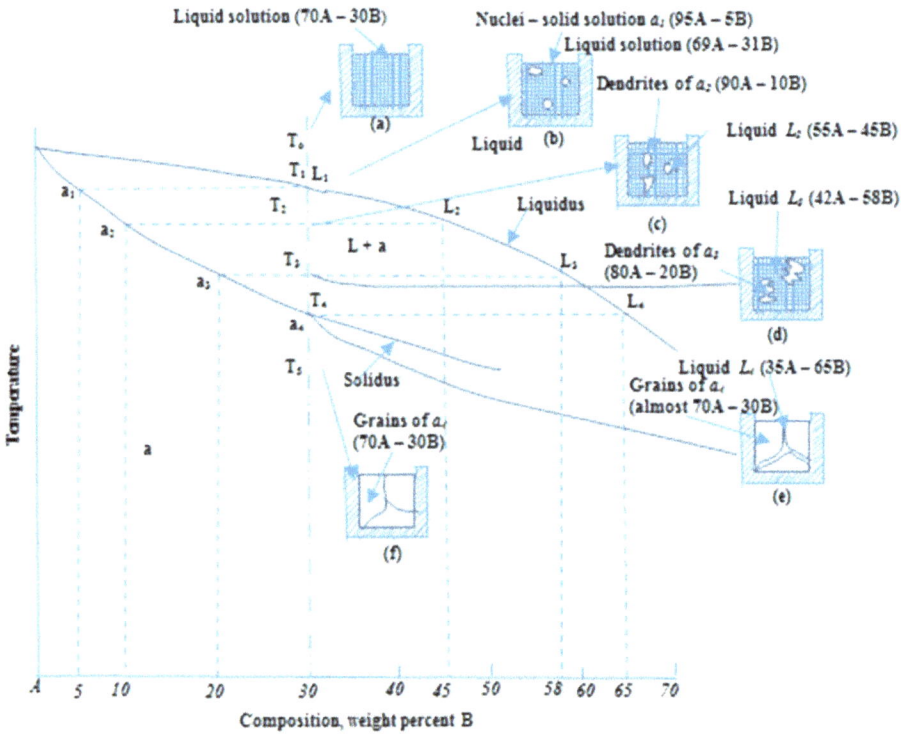

Figure 7.7. Variation of the composition of liquid and solid phases during equilibrium cooling of an alloy (70A–30B). Reproduced with permission from Avner (1997).

Equilibrium cooling

Figure 7.7 shows the very slow cooling, under equilibrium conditions, of a particular alloy 70A–30B, which will now be studied to observe the phase changes that occur. This alloy at temperature T_0 is a homogeneous single-phase liquid solution (a) and

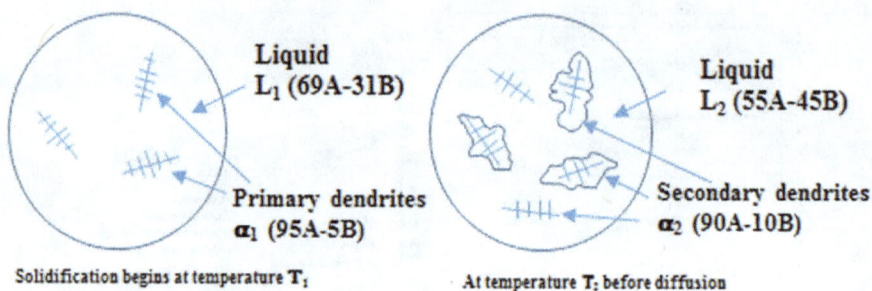

Figure 7.8. Variation of compositions of the liquid and solid phases at T_1 and T_2. Reproduced with permission from Avner (1997).

remains so until temperature T_1 is reached. Since T_1 is on the liquidus line, freezing or solidification now begins. The first nuclei of solid solution to form, α_1, will be very rich in the higher melting point metal A and will be composed of 95A–5B (by the tie line rule). Since the solid solution in forming takes material very rich in A from the liquid, the liquid must become richer in B. Just after the start of solidification, the composition of the liquid is approximated as 69A–31B (b). When the lower temperature T_2 is reached, the liquid composition is at L_2. The only solid solution in equilibrium with L_2 and therefore the only solid solution forming at T_2 is α_2. Applying the tie line rule, α_2 is composed of 10B. Hence, as the temperature is decreased, not only does the liquid composition become richer in B but also the solid solution. At T_2, crystals of α_2 are formed surrounding the α_1 composition cores and also separate dendrites of α_2 (see figure 7.8).

In order for equilibrium to be established at T_2, the entire solid phase must be a composition α_2. This requires diffusion of B atoms to the A-rich core, not only from the solid just formed but also from the liquid. This is possible in the crystal growth (figure 7.7(c)). The composition of the solid solution follows the solidus line while the composition of liquid follows the liquidus line, and both phases are becoming richer in B. At T_3 (d), the solid solution will make up approximately three-fourths of all the material present. Finally, the solidus line is reached at T_4, and the last liquid L_4, very rich in B, solidifies primarily at the grain boundaries (e). However, diffusion will take place and all the solid solution will be of uniform composition α(70A–30B), which is the overall composition of the alloy (f). There are only grains and grain boundaries. There is no evidence of any difference in chemical composition inside the grains, indicating that diffusion has made the grain homogeneous.

Non-equilibrium cooling
In actual practice it is extremely difficult for cooling to occur under equilibrium conditions. Since diffusion in the solid state takes place at a very slow rate, it is expected that with ordinary cooling rates, there will be some difference in the conditions, as indicated by the equilibrium diagram in figure 7.9.

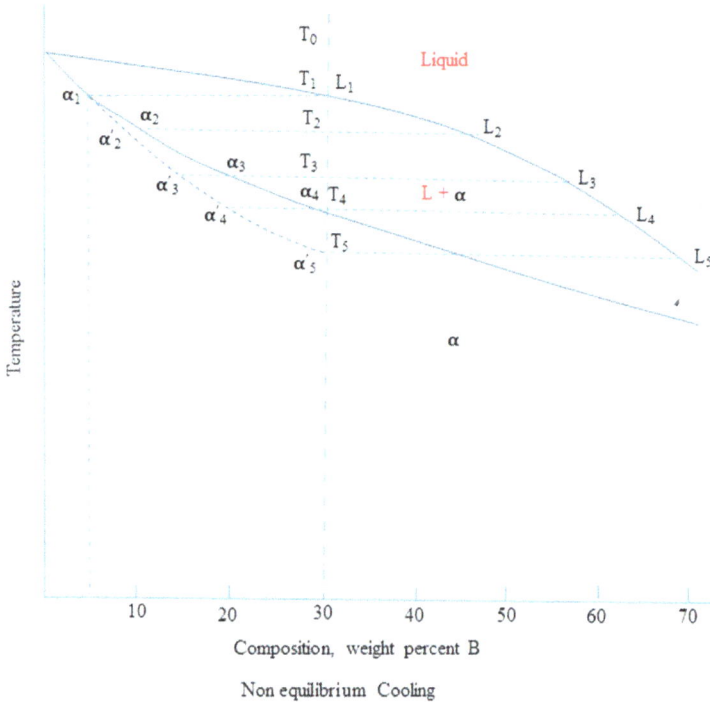

Figure 7.9. Non-equilibrium cooling of an alloy in an isomorphous system. Reproduced with permission from Avner (1997).

Consider the 70A–30B alloy again. Solidification starts at T_1 forming a solid solution of composition α_1. At T_2 the liquid is L_2 and the solid solution now forming is of composition α_2. Since diffusion is too slow to keep pace with crystal growth, not enough time will be allowed to achieve uniformity in the solid, and the average composition will be between α_1 and α_2, say α_2'.

As the temperature decreases, the average composition of the solid solution will depart still further from equilibrium conditions. It seems that the composition of the solid solution is following a 'non-equilibrium' solidus line α_1 to α_5', shown by dotted lines in the figure. The liquid, on the other hand, has essentially the composition given by the liquidus line, since diffusion is relatively rapid in a liquid. At T_3 the average solid solution will be of composition α_3' instead of α_3. Under equilibrium cooling, solidification should be complete at T_4; however, since the average composition of the solid solution α_2' has not reached the composition of the alloy, some liquid must still remain. Applying the lever rule at T_4 gives $\alpha_4' = 75\%$ and $L_4 = 25\%$. Therefore, solidification will continue until T_5 is reached. At this temperature the composition of the solid solution α_5' coincides with the alloy composition, and solidification is complete. The last liquid to solidify, L_5, is richer in B than the last liquid to solidify under equilibrium conditions. The more rapidly the alloy is cooled the greater will be the composition range in the solidified alloy. Since the rate of chemical attack varies with composition, proper etching will reveal the dendritic structure microscopically (see below figure). The final solid consists of a 'cored' structure with a higher-melting central portion surrounded by the lower-melting, last-to-solidify shell. The above condition is referred to as coring or dendritic segregation (see figure 7.10). To summarize, non-equilibrium cooling results in an increased temperature range over which the liquid and solid are present. Since diffusion does not keep pace with crystal growth, there will be a difference in chemical composition from the center to the outside of the grains. The faster the rate of cooling, the greater the above effects will be.

Homogenization

Cored structures are most common in as-cast metals. From the earlier discussion of the origin of a cored structure, it is apparent that the last solid formed along the grain boundaries and in the interdentritic spaces is very rich in the lower-melting-point metal (see figure 7.11). Depending upon the properties of the lower-melting-point

Figure 7.10. Dendritic segregation during solidification. Reproduced from Madison *et al* (2008) with permission of Springer.

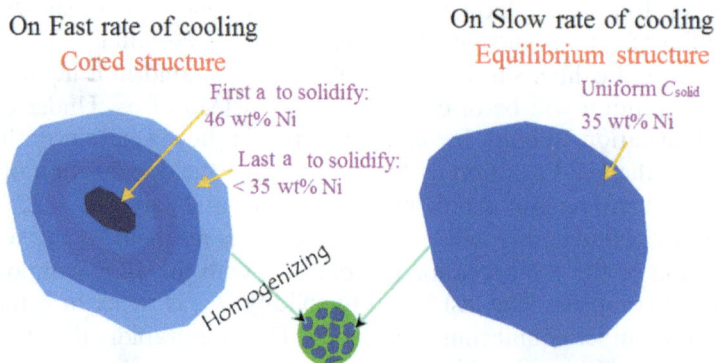

Figure 7.11. (a) Variation of composition in a cored structure and (b) uniformity of composition in a homogenized equilibrium structure.

metal, the grain boundaries may act as a plane of weakness. This will also result in a serious lack of uniformity in the mechanical and physical properties and, in some cases, increased susceptibility to intergranular corrosion because of preferential attack by a corrosive medium. Therefore, for some applications a cored structure is objectionable. There are two methods for solving the problem of coring. One is to prevent their formation by slow freezing from the liquid, but this result in a large grain size and requires a very long time. The preferred method industrially is to achieve equalization of composition or homogenization of the cored structure by diffusion in the solid state. At room temperature, for most metals, the diffusion rate is very slow; but if the alloy is reheated to a temperature below the solidus line, diffusion will be more rapid and homogenization will occur in a relatively short time.

7.4.2 Eutectic phase diagrams

Very few systems exhibit an isomorphous phase diagram (usually the solid solubility of one component in another is limited). Often the solid solubility is severely limited, although the solid solubility is never zero (due to entropic reasons). In a simple eutectic system (binary), there is one composition at which the liquid freezes at a single temperature. This is in some sense similar to a pure solid which freezes at a single temperature (unlike a pure substance the freezing produces two solid phases, both of which contain both components). The term eutectic means easy melting, and an alloy of eutectic composition freezes at a lower temperature than the melting point of the constituent components. This has important implications, e.g. the Pb–Sn eutectic alloy melts at 183 °C, which is lower than the melting points of both Pb (327 °C) and Sn (232 °C). It can be used for soldering purposes (as we want to input, the least possible amount of heat to solder two materials). In figure 7.12 we consider the Pb–Sn eutectic phase diagram.

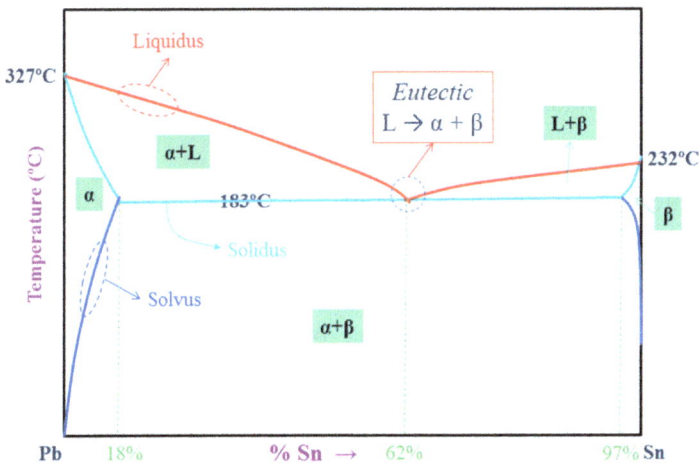

Figure 7.12. Eutectic phase diagram of the Pb–Sn phase diagram.

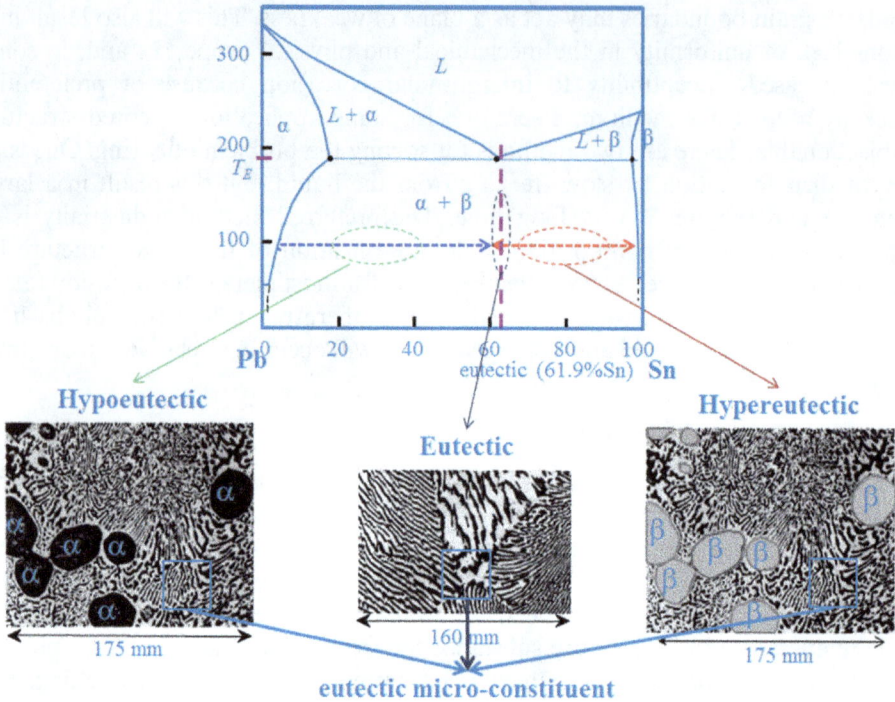

Figure 7.13. Microstructures of hypo- and hyper-eutectic structures for Pb–Sn alloy system.

Microstructural characteristics of eutectic system

To reiterate an important point: phase diagrams do not contain microstructural information (i.e. they cannot tell you what microstructures are produced by cooling). Often microstructural information is overlaid on the phase diagram for convenience. Hence, cooling is strictly not in the domain of the phase diagram, but we can overlay such information keeping in view the assumptions involved. The microstructures at the eutectic, hypo-eutectic and hyper-eutectic regions of the Pb–Sn phase diagram (see figure 7.13).

Special eutectic systems

Consider a eutectic system without terminal solid solubility, Bi–Cd (see figure 7.14). Technically, it is incorrect to draw eutectic phase diagrams with zero solid solubility. This would imply that a pure component (say Bi in the example considered) melts over a range of temperatures (from 'p' to 271 °C), which is an incorrect assumption. Also let us consider an example of a point 'p' (which lies on the 'eutectic line' PQ). At 'p' the phase rule becomes: $F = C - P + 1 = 1 - 3 + 1 = -1$. Note that the above is an alternative way of arriving at the obvious contradiction that at 'p', on the one hand, we are saying that there is a pure component and, on the other hand, we are considering a three-phase equilibrium (which can happen only for Bi–Cd alloys). The correct version of the diagram should have some terminal solid solubility.

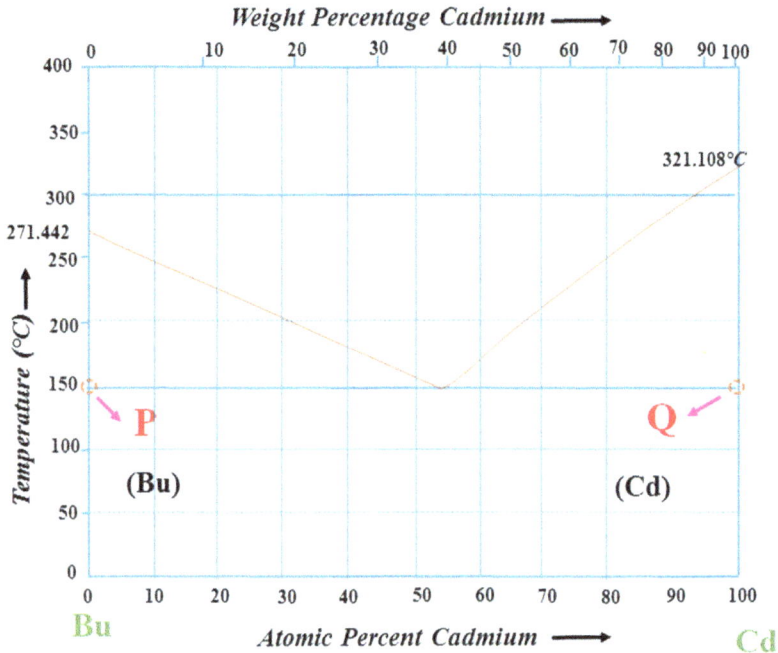

Figure 7.14. Eutectic phase diagram of a Bi–Cd alloy system. According to this phase diagram Bi and Cd have no terminal solid solubility. Reproduced from Moser *et al* (1988) with permission.

In eutectic systems, at eutectic point E, three phases co-exist: L, α and β. The number of components in a binary phase diagram is two and the number of degrees of freedom is $F = 0$. This implies that the eutectic point is an invariant point. In a given system it occurs at a fixed composition and temperature.

7.4.3 Peritectic phase diagrams

Like the eutectic system, the peritectic reaction is found in systems with a complete liquid solubility but limited solid solubility. In the peritectic reaction the liquid (L) reacts with one solid (α) to produce another solid (β): $L + \alpha \rightarrow \beta$. Since the solid β forms at the interface between L and α, further reaction is dependent on solid state diffusion (see figure 7.15). Needless to say this becomes the rate limiting step and hence it is difficult to 'equilibrate' peritectic reactions (compared to, say, eutectic reactions). In some peritectic reactions (e.g. the Pt–Ag system) the (pure) β phase is not stable below the peritectic temperature ($T_P = 1186\,^\circ$C for the Pt–Ag system) and splits into a mixture of ($\alpha + \beta$) just below T_P. The peritectic phase diagram for the Pt–Ag alloy system is given in figure 7.16.

7.4.4 Monotectic phase diagrams

In all the types of phase diagram discussed above, it was assumed that there was complete solubility in the liquid state. It is quite possible, however, that over a

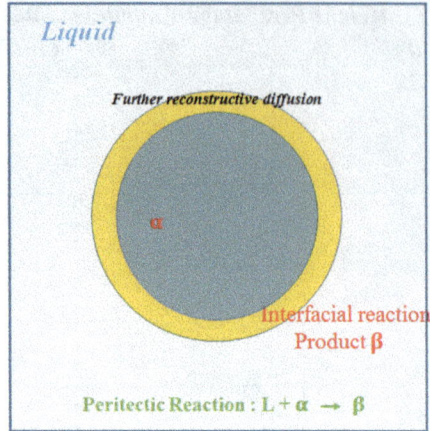

Figure 7.15. Mechanism of the peritectic reaction.

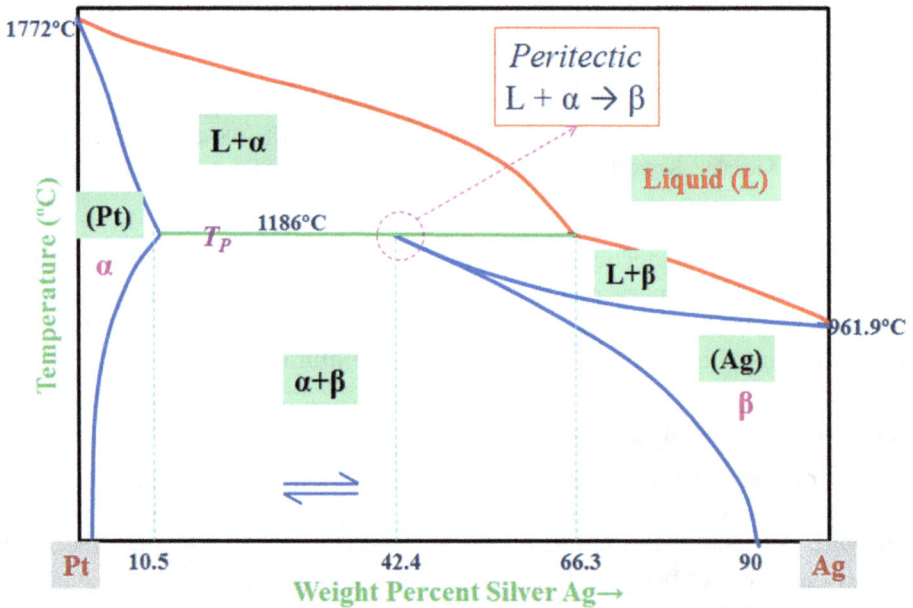

Figure 7.16. Peritectic reaction in the Pt–Ag alloy system. Reproduced with permission from Avner (1997).

certain composition range two liquid solutions are formed that are not soluble in each other. Another term for solubility is miscibility. Substances that are not soluble in each other, such as oil and water, are said to be immiscible. Substances that are partly soluble in each other are said to show a miscibility gap, and this is related to monotectic systems. When one liquid forms from another liquid, plus a solid, on cooling, it is known as a monotectic *reaction*. It should be apparent that the monotectic reaction resembles the eutectic reaction, the only difference being that

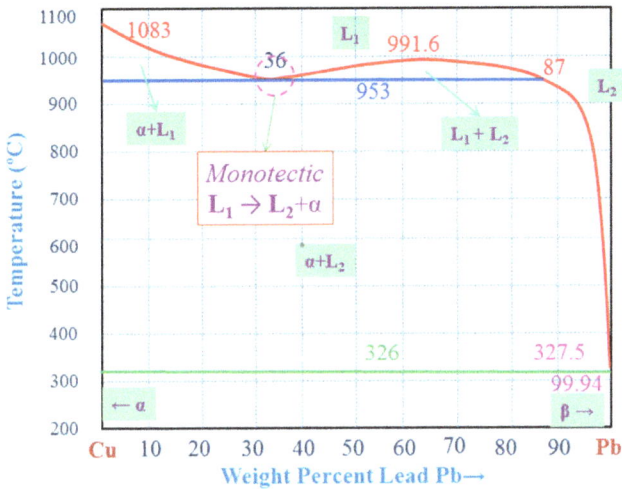

Figure 7.17. Monotectic phase diagram of the Cu–Pb alloy system. Reproduced from American Society for Metals (1948) with permission.

one of the products is a liquid phase instead of a solid phase. An example of an alloy system showing a monotectic reaction is that between copper and lead given below. Note that in this case $L_1 + L_2$ is considered. Also, although the terminal solids are indicated as α and β, the solubility is usually so small that they are practically the pure metals, copper and lead. The monotectic phase diagram for Cu–Pb is given in figure 7.17.

7.4.5 Eutectoid phase diagrams

The eutectoid reaction is a common reaction in the solid state. It is very similar to the eutectic reaction but does not involve the liquid. In this case, a solid phase transforms on cooling into two new solid phases.

The resultant eutectoid mixture is extremely fine, just like the eutectic mixture. Under the microscope both mixtures generally appear the same, and it is not possible to determine microscopically whether the mixture resulted from a eutectic reaction or eutectoid reaction. An equilibrium diagram of Cu–Zn, illustrating the eutectoid reaction, is shown in figure 7.18. The copper–zinc system contains two terminal solid solutions, i.e. these are the extreme ends of phase diagram α and η, with four intermediate phases called β, γ, δ and ε. The β' phase is termed an ordered solid solution, one in which the copper and zinc atoms are situated in a specific and ordered arrangement within each unit cell. In figure 7.18, some phase boundary lines near the bottom are dashed to indicate that their positions have not been exactly determined. The reason for this is that at low temperatures, the diffusion rates are very slow and inordinately long times are required for the attainment of equilibrium. Again, only single- and two-phase regions are found on the diagram, and we can utilize the lever rule for the computing phase compositions and relative amounts.

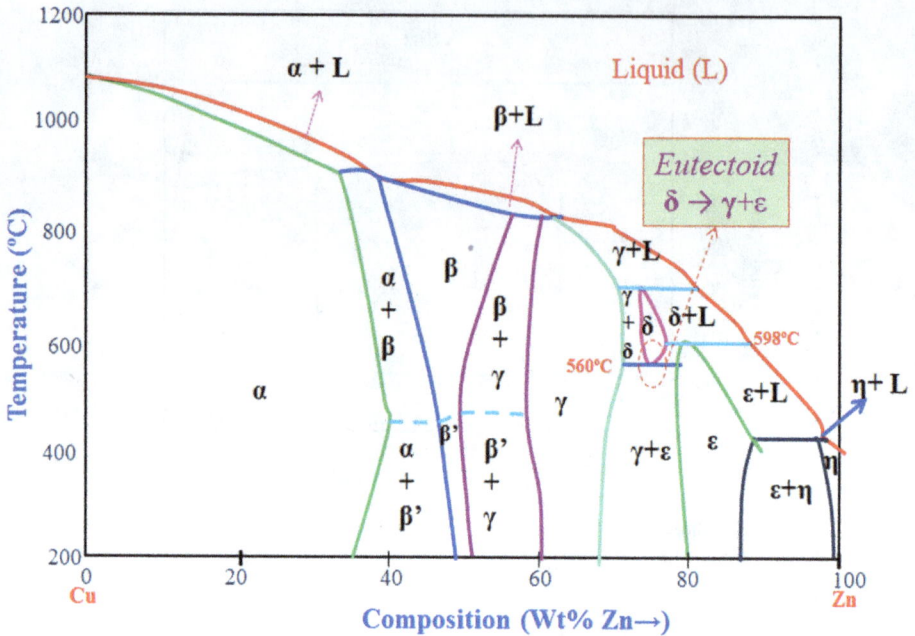

Figure 7.18. Eutectoid transformation in the Cu–Zn alloy system.

The commercial material brasses are copper-rich copper–zinc alloys: for example, cartridge brass has a composition of 70 wt% Cu–30 wt% Zn and a microstructure consisting of a single α phase.

7.4.6 Peritectoid phase diagrams

This is a fairly common reaction in the solid state and appears in many alloy systems. The peritectoid reaction may be written as $solid_1 + solid_2 \leftrightarrow solid_3$. The new solid phase is usually an intermediate alloy, but it may also be a solid solution. The peritectoid reaction has the same relationship to the peritectic reaction as the eutectoid has to the eutectic. Essentially, it is the replacement of a liquid by a solid. The peritectoid reaction occurs entirely in the solid state and usually at lower temperatures than the peritectic reaction, the diffusion rate will be slower and there is less likelihood that equilibrium structures will be reached. Consider the silver–aluminium phase diagram (see figure 7.19) containing a peritectoid reaction. If a 7% Al alloy is rapidly cooled from the two-phase area just above the peritectoid temperature the two phases will be retained, and the microstructure will show a matrix of γ with just a few particles of α. When we cool at below the peritectoid temperature by holding we obtain the single phase μ.

7.4.7 Monotectoid phase diagrams

The monotectoid reaction can also occur in binary systems when neither component shows allotropic transformations; for example, the Al–Zn alloy system, it shows a

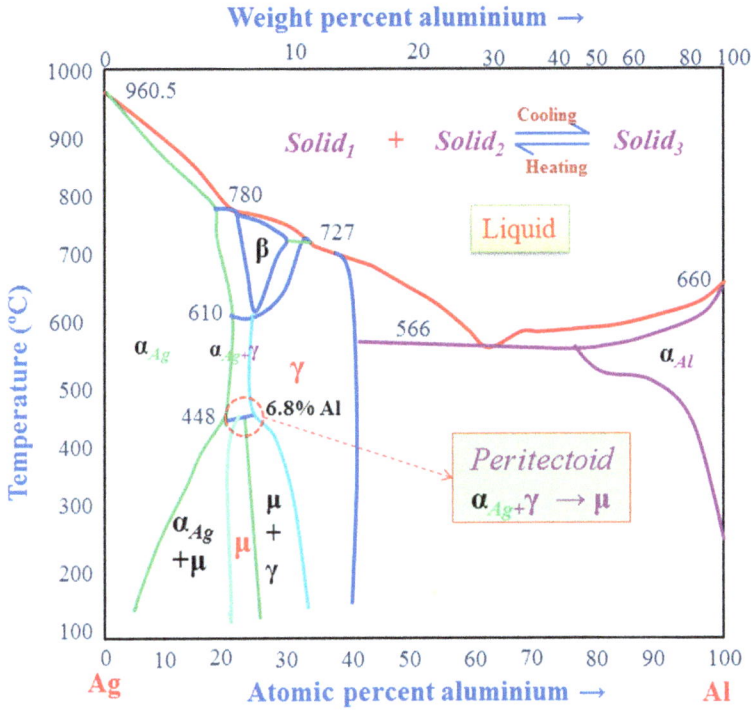

Figure 7.19. Peritectoid reaction in the Ag–Al alloy system.

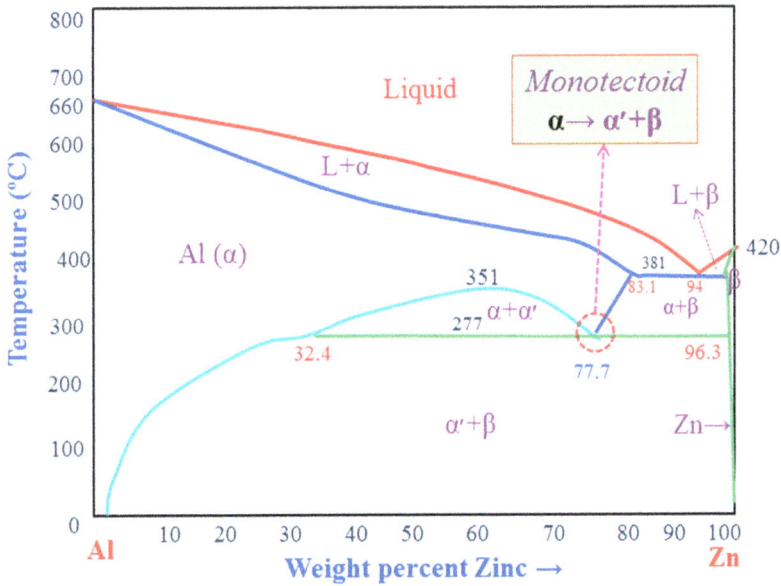

Figure 7.20. Monotectoid reaction in the Al–Zn alloy system.

maximum solubility of 67 atm% Zn in Al at the eutectic temperature. The phase is not stable below 351 °C as a homogeneous solid solution in the whole range of compositions. The following reaction is observed at the monotectoid reaction isotherm, $\alpha \leftrightarrow \beta + \alpha'$ (see figure 7.20).

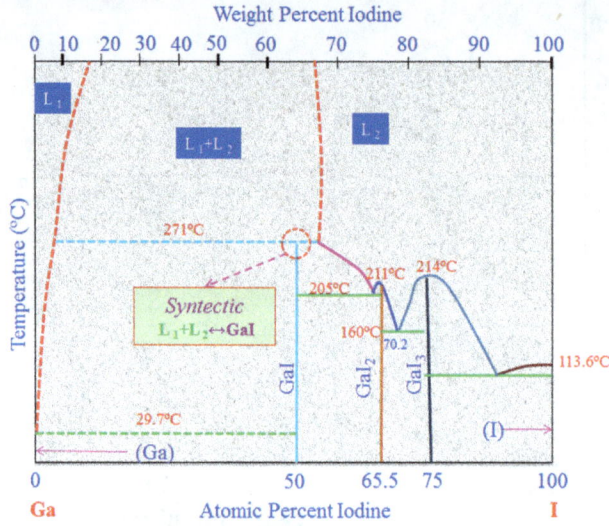

Figure 7.21. Syntectic reaction in the Ga–I phase diagram.

Table 7.2. Summary of invariant reactions.

Name of the reaction	Phase equilibrium	Schematic representation		
Eutectic	$L \leftrightarrow S_1 + S_2$			
Peritectic	$S_1 + L \leftrightarrow S_2$			
Monotectic	$L_1 \leftrightarrow S_1 + L_2$			
Eutectoid	$S_1 \leftrightarrow S_2 + S_3$			
Peritectoid	$S_1 + S_2 \leftrightarrow S_3$			
Monotectoid	$S_{1a} \leftrightarrow S_{1b} + S_2$			
Metatectic	$S_1 \leftrightarrow S_2 + L$			
Syntectic	$L_1 + L_2 \leftrightarrow S$			

The crystal structures of α and a' conform to the fcc structure of pure Al and that of the β phase with the hcp structure of pure Zn. The products of the monotectoid reaction generally appear as a duplex structure.

7.4.8 Syntectic phase diagrams

Another notable invariant reaction that is associated with liquid immiscibility is the syntectic reaction in which two liquid phases react to form a solid phase. The example of the syntactic reaction is the gallium (Ga)–indium (I) phase diagram (see figure 7.21).

All the invariant reactions are summarized in table 7.2 showing both the symbolic reaction and the schematic part of the phase diagram.

Further reading

American Ceramic Society 1964–1993 *Phase Equilibria Diagrams (Phase Diagrams for Ceramists* vols 1–11) (Columbus, OH: American Ceramic Society)

American Society for Metals 1948 *Metals Handbook* (Metals Park, OH: American Society for Metals)

Avner S H 1997 *Introduction to Physical Metallurgy* (New York: McGraw-Hill)

Bhadeshia H K D H and Honeycombe R 2006 *Steel: Microstructure and Properties* 3rd edn (Elsevier)

Kingery W D, Bowen H K and Uhlmann D R 1976 *Introduction to Ceramics* (New York: Wiley) ch 7

Madison J, Spowart J E, Rowenhorst D J and Pollock T M 2008 *JOM* **60** 26

Massalski T B (ed) 1990 *Binary Alloy Phase Diagrams* vols 1–3 (Metals Park, OH: ASM International)

Moser Z *et al* 1988 *Bulletin of Alloy Phase Diagrams* (Berlin: Springer)

Porter D A, Eastererling K E and Sherif M Y 2009 *Phase Ttransformations in Metals and Alloys* (Boca Raton, FL: CRC Press Taylor and Francis Group)

Raghavan V 2004 *Materials Science and Engineering* 5th edn (Englewood Cliffs, NJ: Prentice-Hall)

Reed-Hill R E 1964 *Physical Metallurgy Principles* (New York: Van Nostrand Reinhold)

Rhines F N 1956 *Phase Diagrams in Metallurgy* (New York: McGraw-Hill)

Subramaniam A and Balani K (IITK) *Materials Science and Engineering* (e-book) MHRD, India

Chapter 8

Physical metallurgy of ferrous alloys

8.1 The Fe–Fe$_3$C system

8.1.1 History

The iron–carbon (Fe–C) diagram (see figure 8.1) was the first phase diagram of any alloy to be established, in 1898 by Roberts-Austen, after whom austenite came to be named.

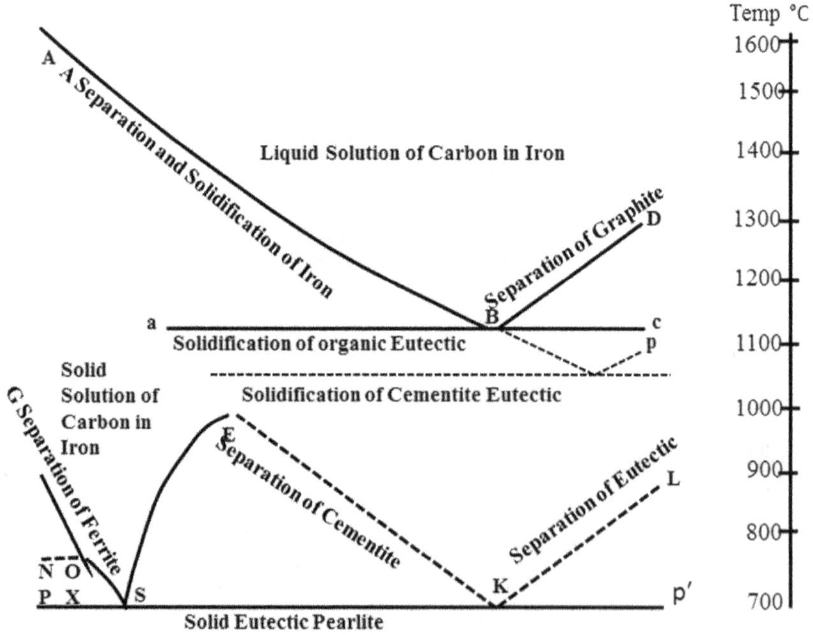

Figure 8.1. Fe–E diagram by Roberts–Austen (1897).

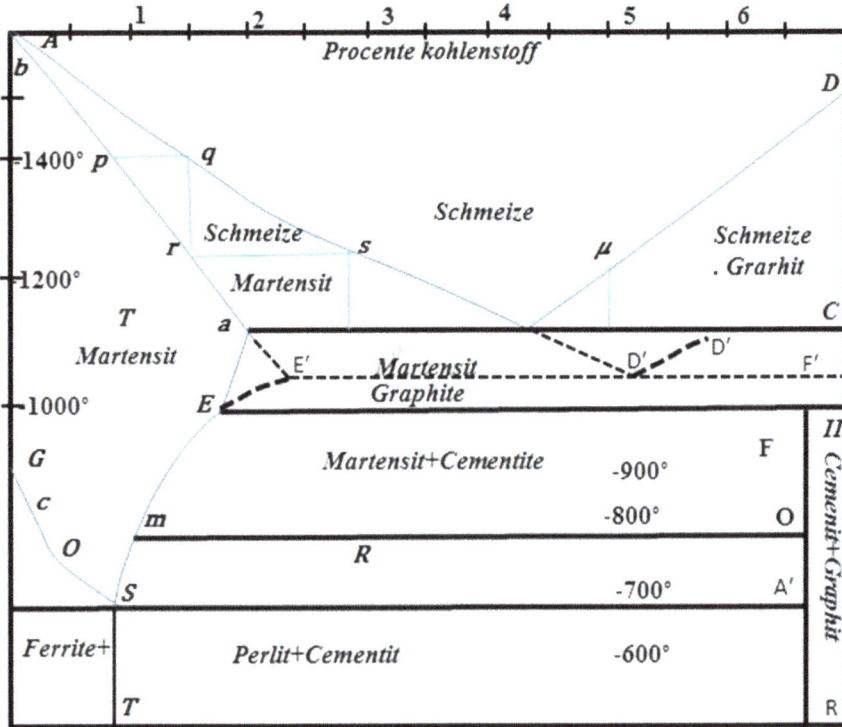

Figure 8.2. Fe–C diagram by Bakhuis–Roozeboom (1900).

The iron–carbon diagram was adjusted by Bakhuis–Roozeboom in 1900 (see figure 8.2). Notice line A–a and the temperature range between 1000 °C and 1100 °C, where carbide formation was supposed to take place as the result of a chemical reaction between graphite and austenite (which at that time was called martensite).

8.1.2 Allotropic transformations in iron

Iron is an allotropic metal, which means that it can exist in more than one type of lattice structure depending on the temperature. A cooling curve for pure iron is shown below in figure 8.3.

There are three more allotropes for pure iron that form under different conditions:

- ε-iron has an hcp structure. This forms at extreme pressure, 110 kbar and 490 °C.
- FCT-iron is fcc tetragonal iron. This is a coherently deposited iron grown as a thin film on a {100} plane of copper substrate.
- Trigonal-iron is iron grown on the misfitting {111} surface of an fcc copper substrate.

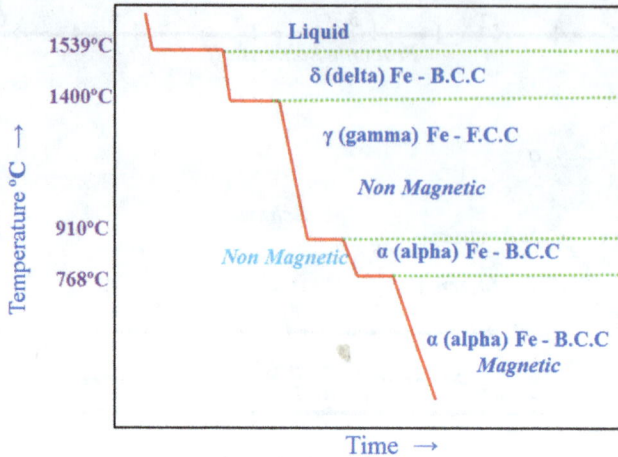

Figure 8.3. Allotropic transformation of pure iron.

8.1.3 The Fe–Fe$_3$C phase diagrams

The Fe–C (or more precisely the Fe–Fe$_3$C) diagram is an important one. Cementite is a metastable phase and, strictly speaking, should not be included in a phase diagram. However, the decomposition rate of cementite is small, and hence it can be thought of as stable enough to be included in a phase diagram. Hence, we typically consider the Fe–Fe$_3$C part of the Fe–C phase diagram. A portion of the Fe–C diagram—the part from pure Fe to 6.67 wt.% carbon (corresponding to cementite, Fe$_3$C)—is technologically very relevant. Cementite is not an equilibrium phase and would tend to decompose into Fe and graphite. This reaction is sluggish and for practical purposes (at the microstructural level) cementite can be considered to be part of the phase diagram. As cementite forms it nucleates readily compared to graphite. Compositions of up to 2.1% C are called steels and beyond 2.1% are called cast irons. In fact, classification should be based on 'castability' and not just on carbon content. Heat treatments can be carried out to modify the properties of the steel by modifying the microstructure. A typical iron–iron carbide phase diagram is given below in figure 8.4.

The steel range is then subdivided by the eutectoid carbon content (0.8% C). Steels containing less than 0.8% C are called hypo-eutectoid steels while those containing between 0.8–2.1% C are called hyper-eutectoid steels. The cast iron range can also be subdivided: eutectic carbon content (4.3% C) cast irons that contain less than 4.3% C are known as hypo-eutectic cast irons, whereas those that contain more than 4.3% C are called hyper-eutectic cast irons.

Why is the concentration of carbon in α-Fe with a bcc structure less than for γ-Fe with an fcc structure? First let us consider the fcc structure (γ-Fe). The packing factor of the fcc lattice is 0.74. This indicates the presence of voids. In an fcc crystal there are mainly two types of voids, tetrahedral and octahedral. Tetrahedral voids have a CN of 4, which means the void is surrounded by four atoms. Octahedral

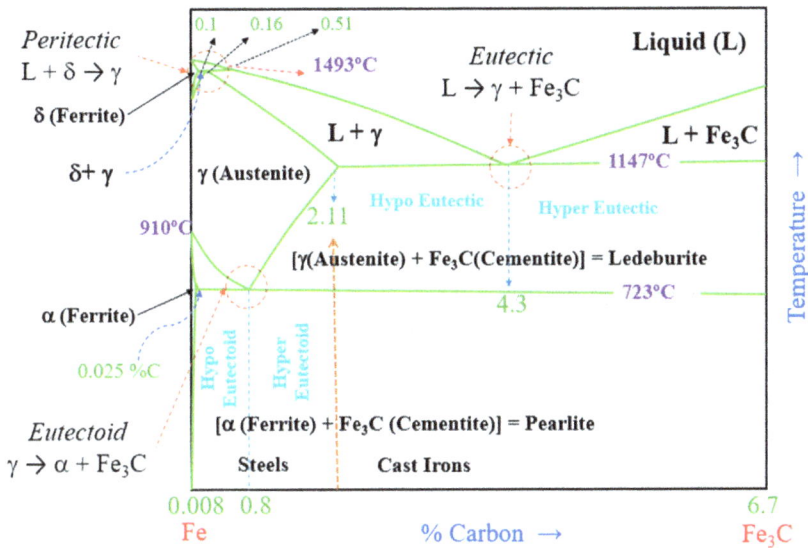

Figure 8.4. Fe–Fe$_3$C phase diagram. Reproduced with permission of Professor Subramaniam, Indian Institute of Technology http://home.iitk.ac.in/~anandh/E-book/.

voids have CN = 6, i.e. the void is surrounded by six atoms. There are eight tetrahedral voids in a unit cell, which means two voids per atom. There are four octahedral voids in a unit cell, which means one void per atom. However, impurities prefer to occupy octahedral voids. The ratio of the radius of the tetrahedral void to the radius of the atom is 0.225 and the same for the octahedral void is 0.414. The ratio of the radius of the carbon atom (77 pm) to that of Fe (when it has fcc crystals) is 0.596. Therefore, when a carbon atom occupies any void, the lattice will be distorted to increase the enthalpy. The distortion will be less if it occupies the octahedral voids. Although they increase the enthalpy, carbon atoms will be present up to a certain extent because of the gain in entropy, as explained previously, which decreases the free energy.

Why is the concentration of carbon in α-Fe with bcc structure less than for γ-Fe with an fcc structure? Consider the bcc unit cell (α-Fe). The packing factor of the bcc lattice is 0.68, thus the total void in a bcc unit cell is higher than in an fcc cell. However, there are 12 (six per atom) tetrahedral and six (three per atom) octahedral voids present. This number is higher than the number of voids present in an fcc unit cell. Unlike voids in an fcc lattice, in a bcc lattice the voids are distorted. This means that if an atom sits in a void, it will not touch all the host atoms. The ratio of the radius of the tetrahedral void to that of the atom is 0.29 and the ratio of the radius of the octahedral void to that of the atom is 0.155. The ratio of the radius of the C atom (77 pm) to Fe (when it has bcc crystals) is 0.612. Therefore, it is expected that in a bcc unit cell, impurities should prefer tetrahedral voids. However, although the octahedral void size is small, the planar radius which has four atoms on the same plane is 79.6 pm, which is larger than the size of the C atom. This means it only needs to distort two other atoms. On the other hand if C sits in the tetrahedral void it has to distort all four atoms. Thus, in α-Fe with a bcc unit cell, C occupies the octahedral voids. Now the octahedral void size in γ-Fe (fcc) is higher than in α-Fe (bcc). So naturally the distortion in a bcc cell will be higher and the activation energy for impurities to occupy a void in a bcc cell will also be higher. This is why we find a much lower solubility limit of C in α-Fe.

8.1.4 The characteristics of phases in the Fe–Fe$_3$C phase diagram

Ferrite (α)
Ferrite is an interstitial solid solution of a small amount of carbon dissolved in α-Fe. The maximum solubility is 0.025% C at 723 °C and it dissolves only 0.008% C at room temperature. It is the softest structure that appears on the diagram. The crystal structure of ferrite (α) is bcc, the tensile strength is 40 000 psi or 275 MPa, elongation is 40% in 2 inches and hardness is <90 HRB. Figure 8.5 shows the microstructure of ferrite.

Cementite (Fe$_3$C)
Cementite or iron carbide, chemical formula Fe$_3$C, contains 6.67% C by weight and it is a metastable phase. It is typically a hard and brittle interstitial compound of low tensile strength (<5000 psi) but high compressive strength. It is the hardest structure that appears on the diagram. Its crystal structure is orthorhombic (see figure 8.6).

Pearlite (α + Fe$_3$C)
Pearlite is a eutectoid mixture containing 0.80% C and is formed at 723 °C on very slow cooling. It is a very fine plate-like or a lamellar mixture of ferrite and cementite. The fine fingerprint mixture called pearlite is shown in figure 8.7. The tensile strength is 120 000 psi or 825 MPa, elongation is 20% in 2 inches and hardness is HRC 20, HRB 95–100, or BHN 250–300. The pearlite is not a phase but the combination of two phases (ferrite + cementite).

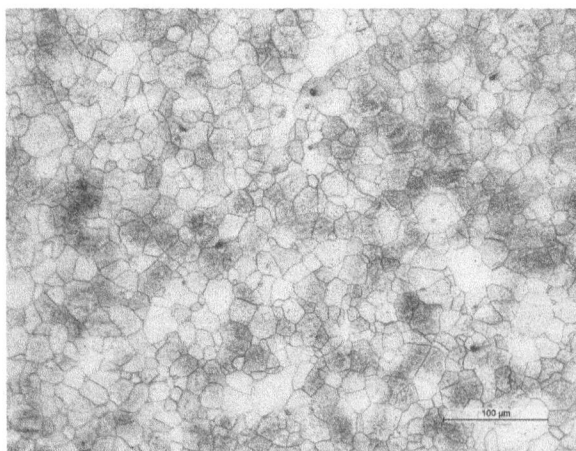

Figure 8.5. Microstructure of α-Fe.

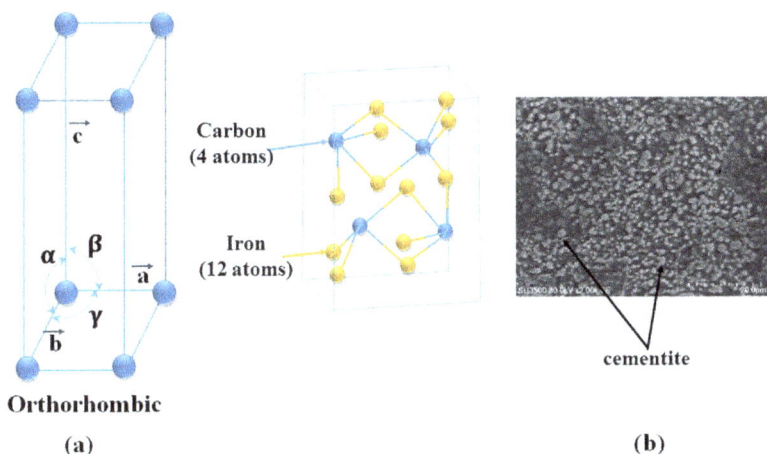

Figure 8.6. (a) Structure and atomic distribution of cementite and (b) microstructure showing pearlite with cementite spheroids.

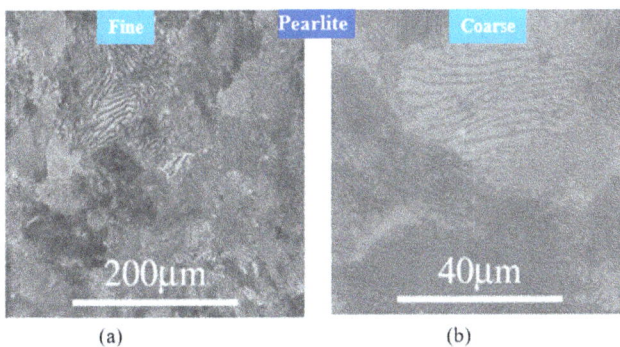

Figure 8.7. Microstructure showing (a) fine pearlite and (b) coarse pearlite. Copyright DoITPoMS Micrograph Library, University of Cambridge. Reproduced with permission.

Austenite (γ)

Austenite is an interstitial solid solution of a small amount of carbon dissolved in γ-Fe. Its maximum solubility is 2.1% C at 1147 °C. The crystal structure of austenite (γ) is fcc. The tensile strength is 150 000 psi or 1035 MPa, elongation 10% in 2 inches, hardness 40 HRC and its toughness are high. A micrograph showing austenite grains is given in figure 8.8, in which twin structures can be observed.

Ledeburite (γ + Fe₃C)

Ledeburite is a eutectic mixture of austenite and cementite. It contains 4.3% C and is formed at 1147 °C. The structure of ledeburite consists of small islands of austenite dispersed in the carbide phase. Ledeburite is not stable at room temperature. The structure of ledeburite is given in figure 8.9.

Figure 8.8. Austenitic grains with twins.

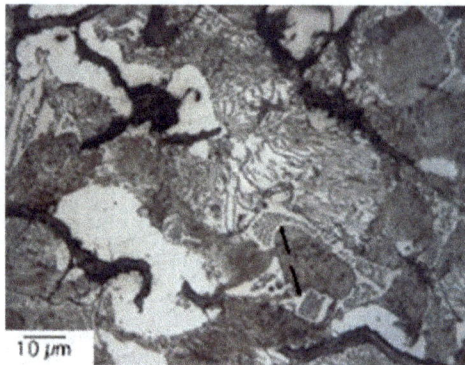

Figure 8.9. Microstructure of ledeburite. Copyright DoITPoMS Micrograph Library, University of Cambridge. Reproduced with permission.

δ-*Ferrite*

δ-Ferrite is an interstitial solid solution of carbon in iron with a bcc crystal structure. δ-Fe has a higher lattice parameter (2.89 Å), compared to austenite, and a solubility limit of 0.09 wt% at 1495 °C. The stability of the phase ranges between 1394 °C and 1539 °C. This material is not stable at room temperature in plain carbon steel. However, it can be present at room temperature in alloy steel, particularly in duplex stainless steel. The microstructure of δ-Fe is given in figure 8.10.

8.1.5 The eutectoid region

These are the phase changes that occur upon passing from the γ region into the α + Fe$_3$C phase field. Consider, for example, an alloy of eutectoid composition (0.8% C) as it is cooled from a temperature within the γ-phase region, say 800 °C— that is, beginning at point 'a' in figure 8.11 and moving down the vertical xx'. Initially the alloy is composed entirely of the austenite phase with composition 0.8 wt.% C and

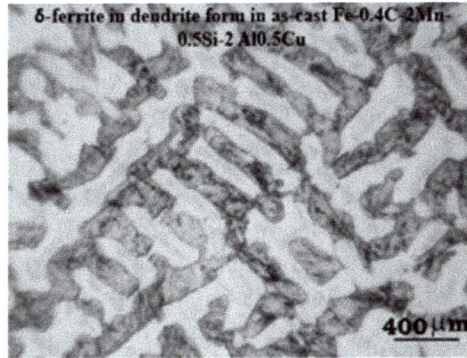

Figure 8.10. Microstructure of δ-ferrite. Copyright DoITPoMS Micrograph Library, University of Cambridge. Reproduced with permission.

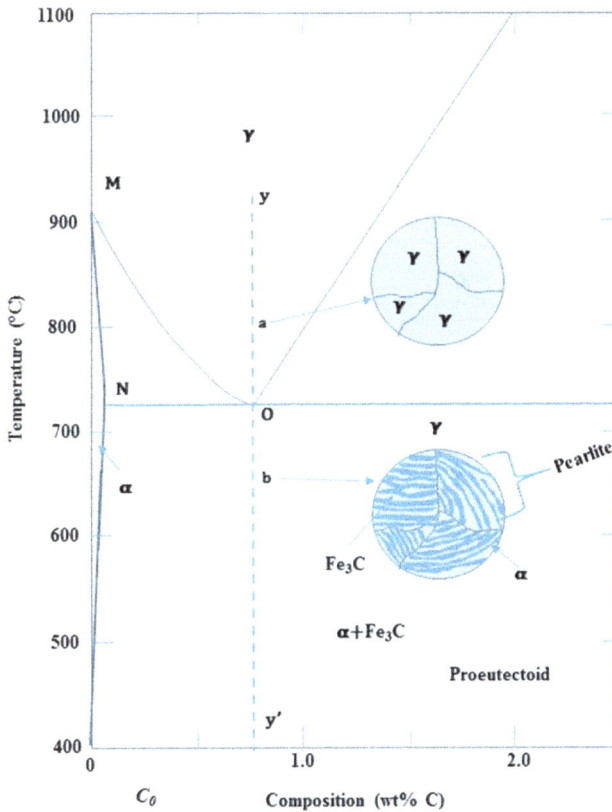

Figure 8.11. Schematic representations of the microstructures for an iron–iron carbon alloy of eutectoid composition.

is then transformed to α + Fe_3C (pearlite). The microstructure of this eutectoid steel that is slowly cooled through the eutectoid temperature consists of alternating layers or lamellae of the two phases α and Fe_3C. The pearlite exists as grains, often termed 'colonies'; within each colony the layers are oriented in essentially the same direction, which varies from one colony to another. The thick light layers are the ferrite phase and the cementite phase appears as thin lamellae, most of which appear dark.

8.1.6 The hypo-eutectoid region

The hypo-eutectoid region corresponds to 0.008–0.8% C. Consider the vertical line yy' in figure 8.12, at about 875 °C, point c, the microstructure will consist entirely of grains of the γ phase. In cooling to point d, about 775 °C, which is within the $\alpha + \gamma$ phase region, both these phases will coexist as in the schematic microstructure. Most of the small α particles will form along the original γ grain boundaries. Cooling from point d to e, just above the eutectoid but still in the $\alpha + \gamma$ region, will produce an increased fraction of the α phase and a microstructure similar to that is also shown; the α particles will have grown larger.

8.1.7 The hyper-eutectoid region

The hyper-eutectoid region is from 0.8–2.1% C. Consider an alloy of composition C_1 in figure 8.13 that, upon cooling, moves down the line zz'. At point g only the γ phase will be present and the microstructure will have only γ grains. Upon cooling into the

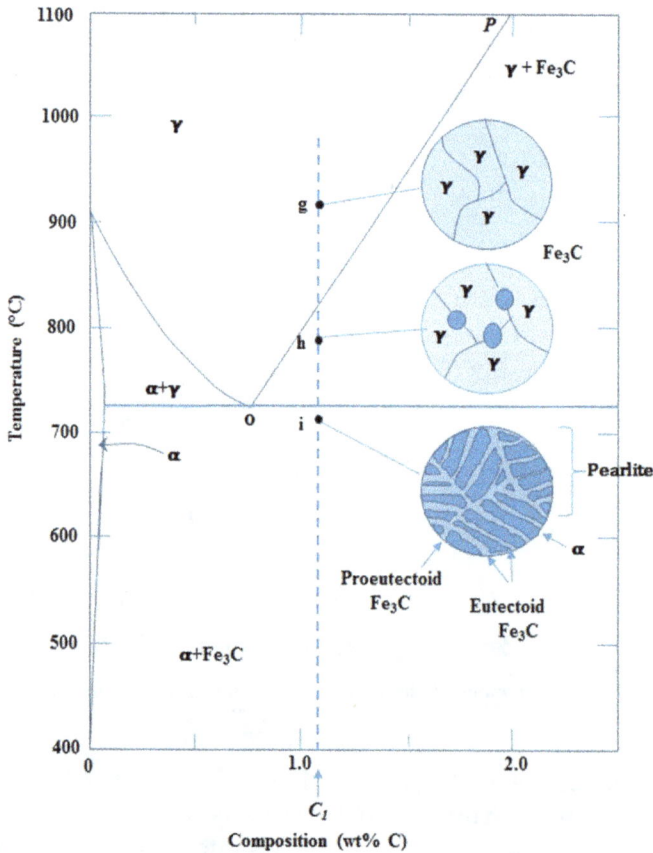

Figure 8.12. Evolution of microstructure during cooling in the hypo-eutectic region.

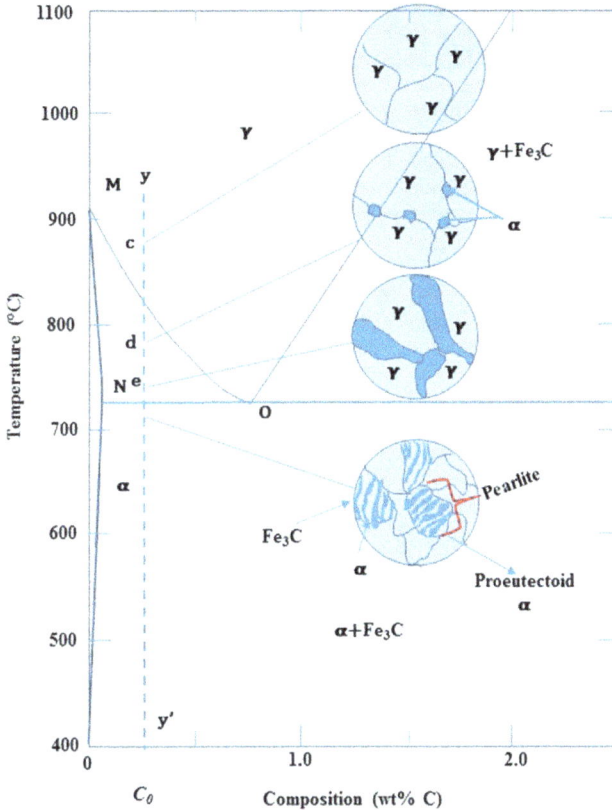

Figure 8.13. Microstructural evolution in the hyper-eutectoid composition range.

γ + Fe$_3$C phase field—say to point h—the cementite phase will begin to form along the initial γ grain boundaries, similar to the α phase at point d. This cementite is called *pro-eutectoid cementite* and forms before the eutectoid reaction. As the temperature is lowered through the eutectoid to point i, all the remaining austenite of eutectoid composition is converted into pearlite; thus the resulting microstructure consists of pearlite and pro-eutectoid cementite as microconstituents.

Example: For a 99.6 wt.% Fe–0.40 wt.% of carbon at a temperature just below the eutectoid, determine the following:
  (a) The amount of Fe$_3$C, ferrite, ferrite (α) and pearlite.
  (b) The amount of pearlite and pro-eutectoid ferrite (α).

Solution: (a) The amount of Fe$_3$C and ferrite (α):

Consider figure 8.14(a). The percentage of Fe$_3$C $= \frac{0.4 - 0.025}{6.67 - 0.025} * 100$.

The percentage of Fe$_3$C in 0.4% C steel is 5.64%.
The percentage of ferrite (α) in 0.4% C steel = (100 − 5.64)%
The percentage of ferrite in 0.4% C steel = 94.36%.

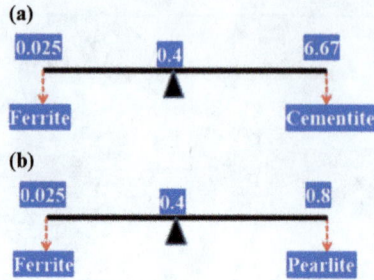

Figure 8.14. (a) Schematic of lever arm of 0.4% C steel for ferrite and cementite. (b) Schematic of lever arm of 0.4% C steel for ferrite and pearlite.

The percentage of ferrite $= \frac{6.67 - 0.4}{6.67 - 0.025} * 100 = 94.36\%$.

(b) The percentage of pearlite and pro-eutectoid ferrite (α): Consider figure 8.14(b)
The percentage of pearlite $= \frac{0.4 - 0.025}{0.8 - 0.025} * 100$.
The percentage of pearlite $= 48\%$.
The percentage of pro-eutectoid ferrite (α) in 0.4% C steel $= (100 - 48)\%$.
The percentage of pro-eutectoid ferrite (α) $= 52\%$.
The percentage of pro-eutectoid ferrite $= \frac{0.8 - 0.4}{0.8 - 0.025} * 100 = 52\%$.

8.1.8 Critical temperatures in the iron–iron carbide phase diagram

In figure 8.15, A_1 is the lower critical temperature, A_3 is the upper critical temperature, A_4 is the eutectic temperature, A_5 is the peritectic temperature and A_{cm} is the

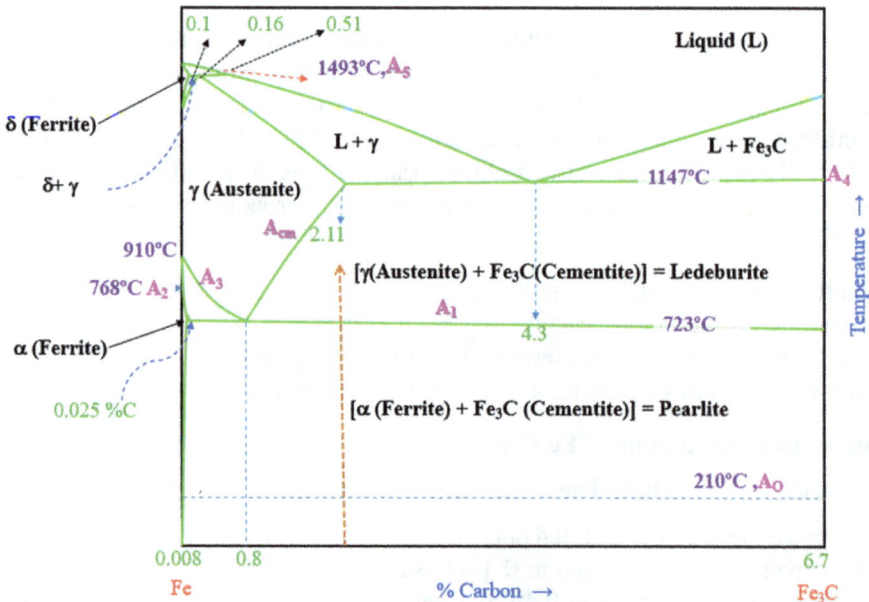

Figure 8.15. Critical temperatures in the Fe–Fe$_3$C phase diagram.

γ/γ + cementite phase field boundary. While heating we denoted these as A_{c1}, A_{c2}, A_{c3}, etc, 'c' stands for *chauffage* (a French word which means heating) and while cooling we denoted them as A_{r1}, A_{r2}, A_{r3}, etc, 'r' stands for *refroidissement* (a French word which means cooling). The upper and lower critical temperature lines are shown as single lines under equilibrium conditions and are sometimes indicated as A_{e3}, A_{e1}, etc. When the critical lines are determined, it is found that they do not occur at the same temperature. The critical line on heating is always higher than the critical line on cooling. Therefore, the upper critical line of a hypo-eutectoid steel on heating would be labeled A_{c3} and the same line on cooling A_{r3}. The rate of heating and cooling has a definite effect on the temperature gap between these lines.

8.1.9 Classification of metal alloys

Figure 8.16 shows a classification of metal alloys.

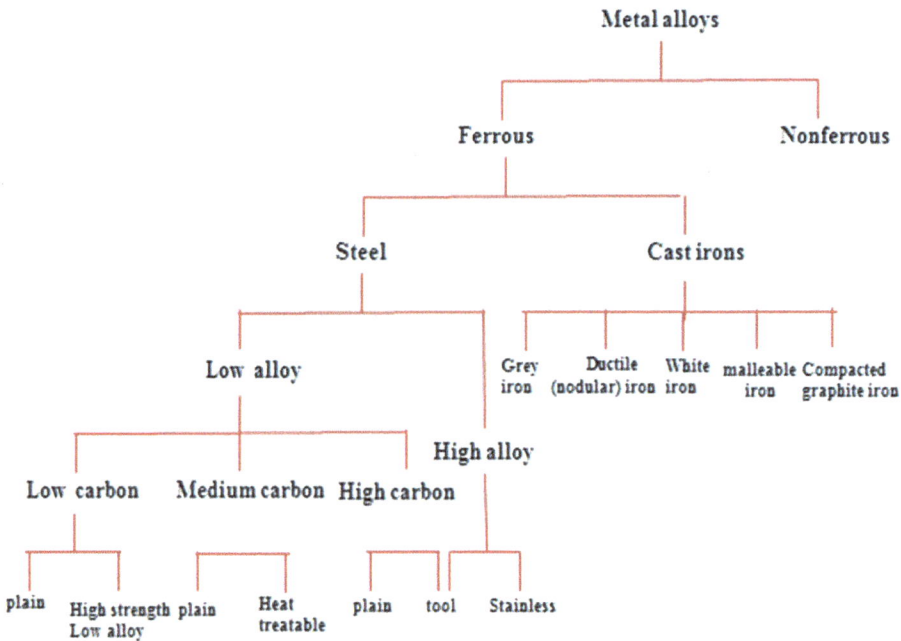

Figure 8.16. Classification of metal alloys.

8.2 Heat treatment of steels

8.2.1 Definition

Heat treatment is an operation or combination of operations involving the heating at a specific rate, soaking at a temperature for a period of time and cooling at some specified rate. The aim is to obtain a desired microstructure in order to achieve certain predetermined properties.

8.2.2 Objectives

The objectives of heat treatment for metals are as follows:
- to increase the strength, hardness and wear resistance;
- to increase ductility, toughness and softness;
- to obtain a fine grain size;
- to remove internal stresses induced by differential deformation by cold working, and non-uniform cooling from high temperatures during casting and welding.

8.2.3 Classification

Figure 8.17 shows the different types of heat treatment processes in steel.

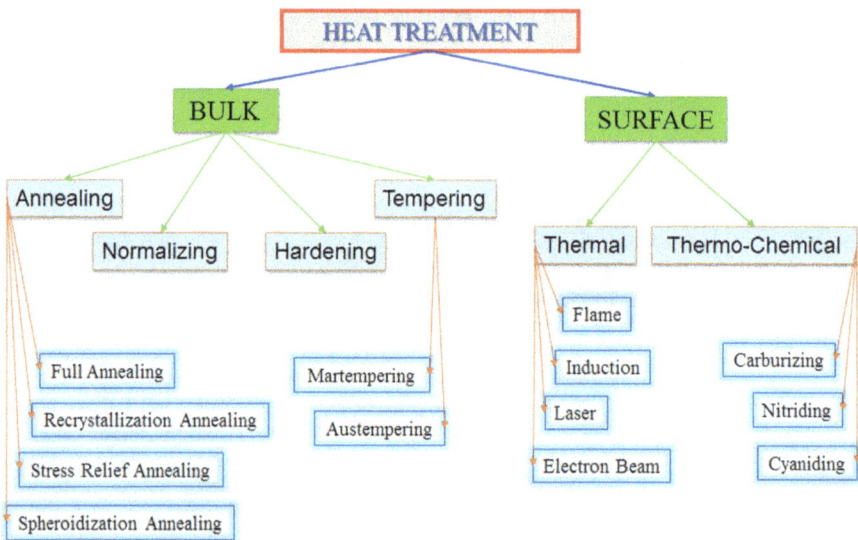

Figure 8.17. Different types of heat treatment processes in steels.

8.2.4 Annealing

Full annealing. The steel is heated above A_3 (for hypo-eutectoid steels) or A_1 (for hyper-eutectoid steels) (see the part of Fe–Fe$_3$C phase diagram, figure 8.18), it is then held at that temperature to obtain the homogeneous austenite phase, then the steel is furnace cooled to obtain coarse pearlite. This coarse pearlite has decreased in hardness and its ductility properties are improved. In the hyper-eutectoid steel heat treatment is carried out below the A_{cm} temperature in order to avoid a continuous network of pro-eutectoid cementite along grain boundaries (crack propagation is aided by this network). The temperature range for carrying out annealing is shown in figure 8.18.

Recrystallization annealing. Heat treatment is carried out below A_1 (see figure 8.18) for a sufficient time to recrystallize the cold worked grains to form new stress-free grains. This method is used between processing steps (e.g. sheet rolling).

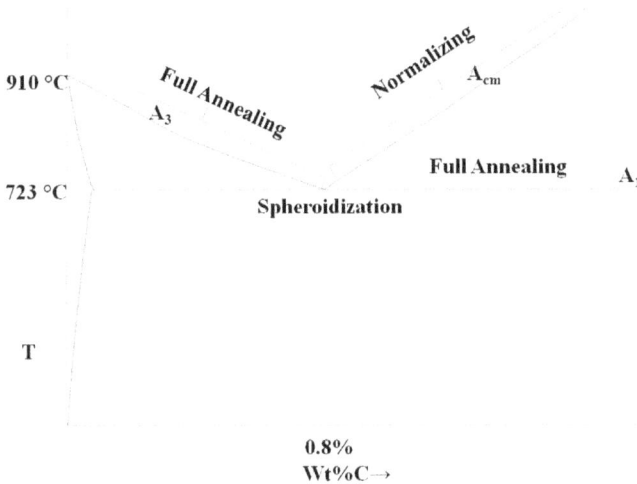

Figure 8.18. Heat treatment temperatures for steels.

Stress relief annealing. Residual stresses in steels appear due to various mechanical operations such as cold rolling/working, machining, differential cooling, the formation of martensite, welding, forging, etc. This residual stress can cause metal failure in various circumstances. The distortion of steel during machining is also caused by the presence of residual stresses. Therefore, the removal of residual stress becomes important for the better performance of the metal. During stress relief annealing the worked metal is heated below the A_1 line for 2 h (see figure 8.18). Residual stress is removed by the annihilation of dislocations and polygonization. Thus stress is removed without introducing any changes in the microstructure and the mechanical properties of the metal are enhanced.

Spheroidization annealing. Ductility in medium- and high-carbon steels is improved by spheroidization annealing. High-carbon steel requiring extensive machining prior to final hardening and tempering is usually subjected to spheroidization annealing to improve ductility. During this process steel is heated above the A_1 line for a prolonged period (see figure 8.18). This causes the shape of cementite plates to change to spheroids, thus enhancing the overall ductility. The driving force for this process is the decrease in the interfacial energy. The spheroidized structure is desirable when minimum hardness, maximum ductility, or (in high-carbon steels) maximum machinability is important. Low-carbon steels are seldom spheroidized for machining, because in the spheroidized condition they are excessively soft and 'gummy'. Medium-carbon steels are sometimes spheroidization annealed to obtain maximum ductility.

8.2.5 Normalizing

During normalizing, hypo-eutectoid steels and hyper-eutectoid steels are heated above the A_3 and A_{cm} lines (50 °C above the annealing temperature; see figure 8.18), respectively, until complete austenization occurs and the steel is then cooled in still air. The formation of fine pearlite in normalizing imparts higher hardness to the steel. The purposes of normalizing are mainly to refine the grain structure prior to hardening, to harden the steel slightly and to reduce segregation in casting or forgings.

8.2.6 Hardening

Hardening involves the heating of the steel above the critical temperature (above A_3 or A_{cm}) for complete austenization, after which it is quenched to room temperature (with a cooling rate higher than the critical cooling rate (CCR)). The temperature zone for the hardening operation is given in figure 8.18. Under slow cooling rates, the carbon atoms are able to diffuse out of the austenite structure and this leads to γ–α transformation. This process involves nucleation and growth and is time-dependent. With a still further increase in cooling rate, insufficient time is allowed for the carbon to diffuse out of the solution, and although some movement of the iron atoms takes place, the structure cannot become bcc while the carbon is trapped in the solution. The resultant structure is called martensite, a supersaturated solid solution of carbon trapped in a body-centered tetragonal structure in a metastable phase. The highly distorted lattice structure is the prime reason for the high hardness of martensite. After drastic cooling, martensite appears microscopically as a white needle-like, acicular or lenticular structures, sometimes described as a pile of straw.

Characteristics of martensite
Martensite is formed by three main mechanisms: Bain distortion, secondary shear rotation and rigid body rotation (see figure 8.19).

Martensitic transformation can be understood by first considering an alternative unit cell for the austenite phase as shown in figure 8.20.

If there is no carbon in the austenite (as in figure 8.20), then the martensitic transformation can be understood as a ~20% contraction along the c-axis and a ~12% expansion of the a-axis, accompanied by no volume change, and the resultant structure has a bcc lattice (the usual bcc-Fe) and a *c/a* ratio of 1.0. In the presence of carbon in the octahedral voids of ccp (fcc) γ-Fe, the contraction along

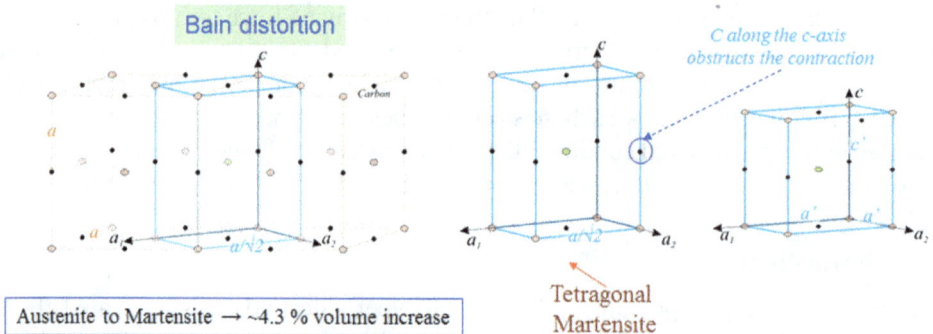

Figure 8.19. Formation of martensite by shear rotation. Reproduced from American Society for Metals (1948) with permission.

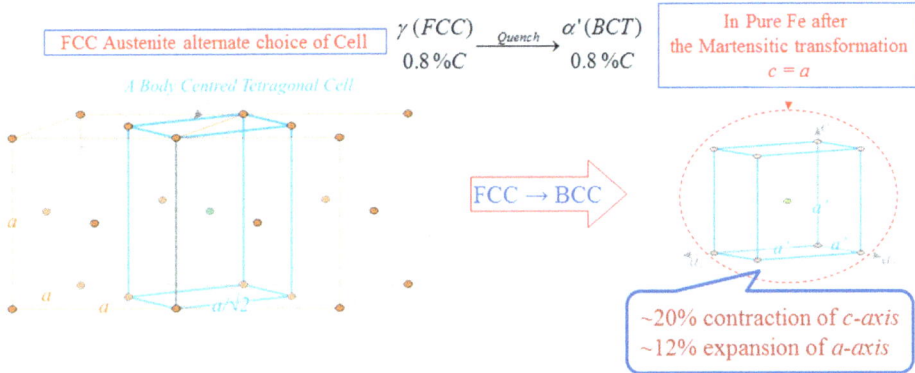

Figure 8.20. Transition of fcc iron to the bcc form (in pure iron). Reproduced with permission of Professor Subramaniam, Indian Institute of Technology http://home.iitk.ac.in/~anandh/E-book/.

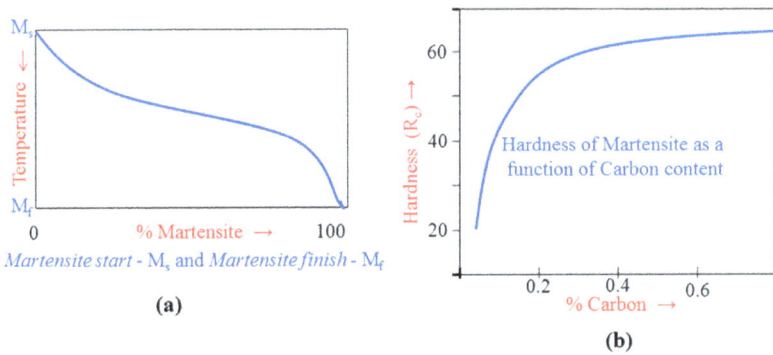

Figure 8.21. (a) Variation of the amount of martensite formed with temperature (from the martensite start temperature to the martensite finish temperature) and (b) hardness with carbon percentage.

the c-axis is impeded by the carbon atoms. (Note that only a fraction of the octahedral voids is filled with carbon as the percentage of C in Fe is small.)

However, the a_1 and a_2 axes can expand freely. This leads to a product with c/a ratio greater than one. In this case there is an overall increase in volume of ~4.3% (depending on the carbon content). This is known as the Bain distortion. Martensitic transformation occurs only when the temperature exceeds a specific value (the martensitic start temperature M_s) during rapid quenching and martensite transformation is completed at a specific temperature known as the martensite finish (M_f) temperature. The hardness of the martensite depends on the carbon percentage. This can be understood from figure 8.21.

Martensite is a supersaturated solid solution of carbon trapped in a body-centered tetragonal structure. Two dimensions of the unit cell are equal, but the third is slightly expanded because of the trapped carbon. The axial c/a ratio increases with carbon content by a maximum of 1.08. This can be seen in figure 8.22. The martensite start and finish temperatures also depend on the carbon content (see figure 8.23).

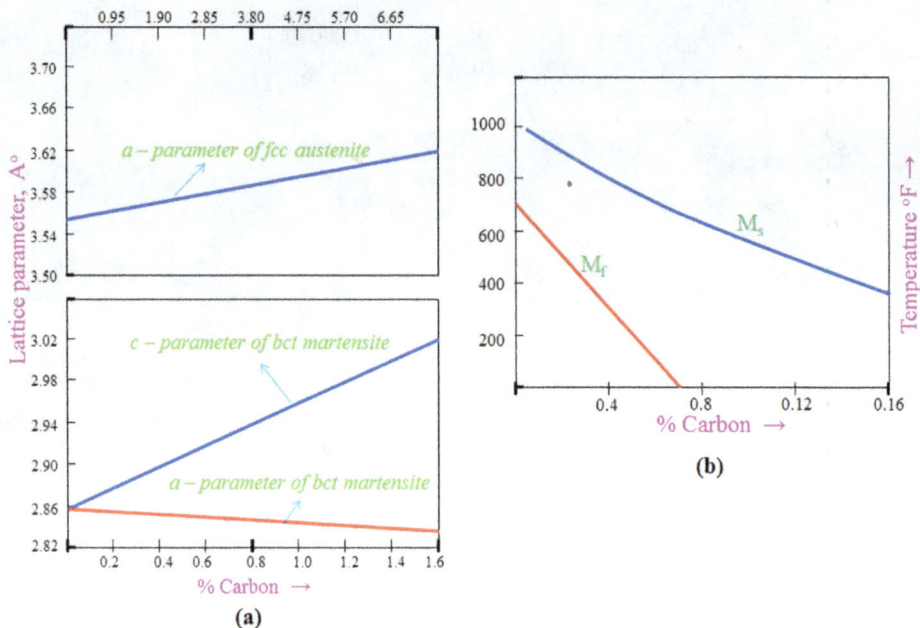

Figure 8.22. (a) Variation of lattice parameters with carbon percentage for fcc and bcc crystals and (b) variation of the martensite start and martensite finish temperatures with carbon percentage.

The martensitic transformation occurs without composition change. The transformation occurs by shear without the need for diffusion and is called diffusionless transformation. The atomic movements (shearing) required are only a fraction of the interatomic spacing.

The amount of martensite formed is a function of the temperature to which the sample is quenched and not of time. The hardness of martensite is a function of the carbon content, but high hardness steel is very brittle and martensite is also brittle. To increase the ductility of hardened steel tempering is performed. The martensite transformation, for many years, was believed to be unique to steel. However, in recent years, this martensite type of transformation has been found in a number of other alloy systems, such as iron–nickel, copper–zinc and copper–aluminum. The basic purpose of hardening is to produce a fully martensitic structure, and the minimum cooling rate (per second) that will avoid the formation of any of the softer products of transformation is known as the critical cooling rate (CCR). The CCR, determined by the chemical composition and austenitic grain size, is an important property of a steel since it indicates how fast a steel must be cooled in order to form only martensite. If the carbon percentage in martensite is less than 4% then the microstructure shows lathe martensite, but if the steel has a carbon content higher than that (4–6%) then plate-like martensite can be observed in the microstructure. The typical microstructure of martensite is given in figure 8.23.

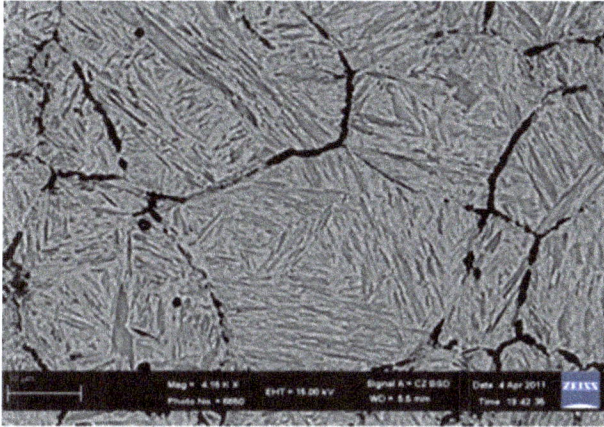

(a)

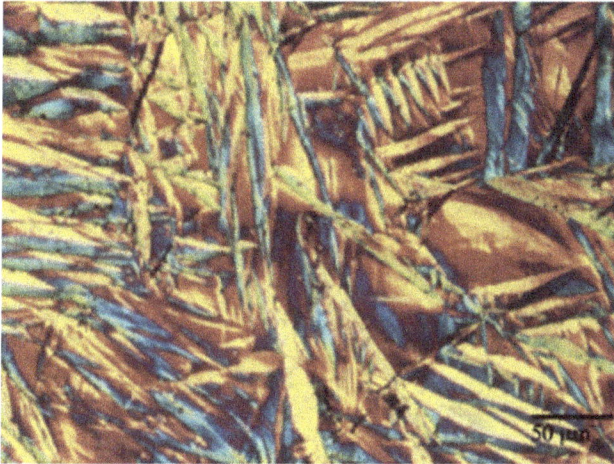

(b)

Figure 8.23. Microstructures of martensite: (a) scanning electron micrograph and (b) optical micrograph.

8.3 Time–temperature transformation diagrams

Davenport and Bain were the first to develop the TTT diagram of eutectoid steel. They determined the pearlite and bainite portions, and Cohen later modified and included the M_s and M_f temperatures for martensite. Some methods are used to determine TTT diagrams. The most popular method is a salt bath technique combined with metallography and hardness measurement. In addition there are other techniques such as dilatometry, electrical resistivity methods, magnetic permeability, *in situ* diffraction techniques (x-ray, neutron), acoustic emission, thermal measurement techniques, density measurement techniques and thermodynamic predictions. TTT diagrams are also called isothermal (constant temperature) transformation diagrams, because TTT diagrams give the kinetics of isothermal transformations. For every composition of steel a different TTT diagram should be drawn.

8.3.1 Determination of the TTT diagram for eutectoid steel

For the determination of isothermal transformation (or TTT) diagrams, we consider a molten salt bath technique combined with metallography and hardness measurements. In a molten salt bath technique, two salt baths and one water bath are used. Salt bath 1 is maintained at the austenizing temperature (780 °C for eutectoid steel). Salt bath 2 is maintained at specified temperature at which the transformation is to be determined (below A_{e1}), typically 700–250 °C for eutectoid steel. Bath 3, which is a cold water bath, is maintained at room temperature. In bath 1 a number of samples are austenized at $A_1 + 20$–40 °C for eutectoid steel, $A_3 + 20$–40 °C for hypo-eutectoid steel and $A_{Cm} + 20$–40 °C for hyper-eutectoid steels, for about one hour. Then samples are removed from bath 1 and placed in bath 2, and each one is kept there for different specified periods of time, say $t_1, t_2, t_3, t_4, ..., t_n$, etc. After the specified times, the samples are removed and quenched in cold water. The microstructure of each sample is studied using metallographic techniques. The type, as well as quantity of phases, is determined for each sample. The TTT diagram for eutectoid steel is given in figure 8.24.

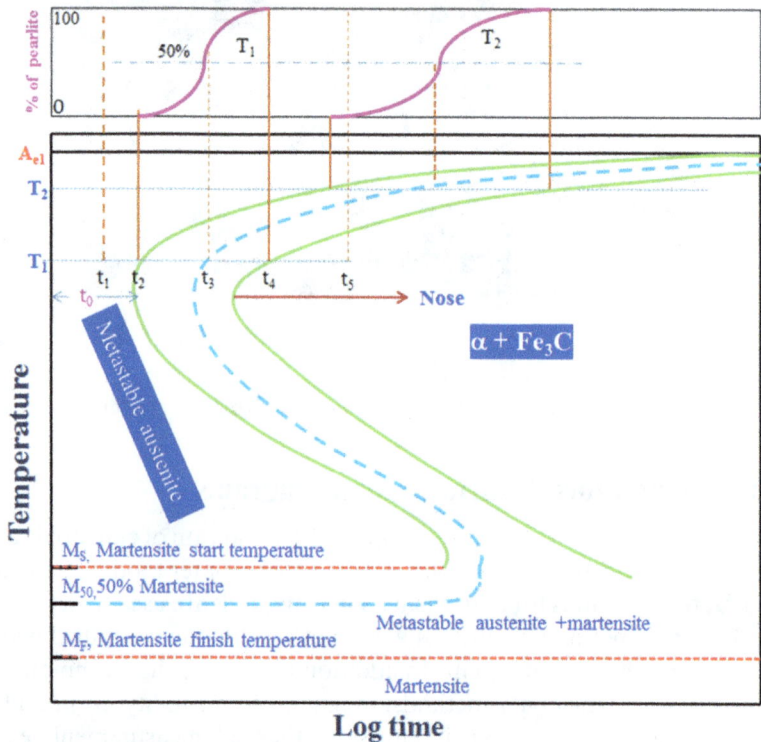

Figure 8.24. TTT diagram for eutectoid steel. Reproduced with permission from *Atlas of Isothermal Transformation Diagrams* (United States Steel Corporation).

Holding the temperature T_1 pearlite formation starts at time t_2 and finishes at time t_4. At t_3 50% of pearlite is formed. For temperature T_2 pearlite formation starts and finishes at a specific time. The x-axis is in the log scale. The 'nose' of the C curve is in seconds and just below T_E the transformation time may be in days. The starting phase (left of the C curve) has to be austenite. To the right of the end of the C curve is the ferrite + Fe_3C phase field. As pointed out above, one of the important utilities of the TTT diagrams comes from the overlay of microconstituents (microstructures) on the diagram. Depending on T, the α + Fe_3C phase field is labeled with microconstituents such as pearlite and bainite. The time taken for 1% transformation to pearlite or bainite is considered as the transformation start time and 99% transformation represents the transformation finish time. We have seen that TTT diagrams are drawn from an instantaneous quench to temperature followed by an isothermal hold. Suppose we quench below $\sim$225 °C, i.e. below the temperature marked M_s, then austenite transforms via a diffusionless transformation (involving shear) to a (hard) phase known as martensite. Below the temperature marked M_f this transformation to martensite is completed. Once γ is exhausted, it cannot transform to α' + Fe_3C. Hence, we have a new phase field for martensite. The fraction of martensite formed is not a function of the time of hold, but the temperature to which we quench (between M_s and M_f). Strictly speaking, cooling curves (including finite quenching rates) should not be overlaid on TTT diagrams (remember that TTT diagrams are drawn for isothermal holds). Generally, two distinct curves exist in TTT diagrams for pearlite and bainite transformations. But they cannot be resolved clearly in plain carbon steels. In alloy steels, these curves are distinct.

8.3.2 TTT diagrams for hypo- and hyper-eutectoid steel

In hypo-eutectoid steels (say composition C_1) there is one more branch to the C curve, NP (see figure 8.25). The part of the curve lying between T_1 and T_E is clear, because in this range of temperatures we expect only pro-eutectoid α to form and the final microstructure will consist of α and γ (e.g. if we cool to T_x and hold).

The part of the curve below T_E is a bit of confusion, since we are instantaneously cooling to below T_E we should obtain a mix of α + Fe_3C. Suppose we quench instantaneously a hypo-eutectoid composition to T_x. We should expect the formation of α + Fe_3C (and not pro-eutectoid α first). The reason we see the formation of pro-eutectoid α first is that the undercooling w.r.t. to A_{cm} is more than the undercooling w.r.t. to A_1. Hence, there is a higher propensity for the formation of pro-eutectoid α.

Similar to the hypo-eutectoid case, the hyper-eutectoid composition (see figure 8.25) will have a γ + Fe_3C branch. For a temperature between T_2 and T_E we end up with γ + Fe_3C. For a temperature below T_E (but above the nose of the C curve), say T_n, first we have the formation of pro-eutectoid Fe_3C followed by the formation of eutectoid α + Fe_3C.

8.3.3 Transformation to pearlite

The transformation product above the nose region is pearlite. The pearlite microstructure is the characteristic lamellar structure of alternate layers of ferrite and

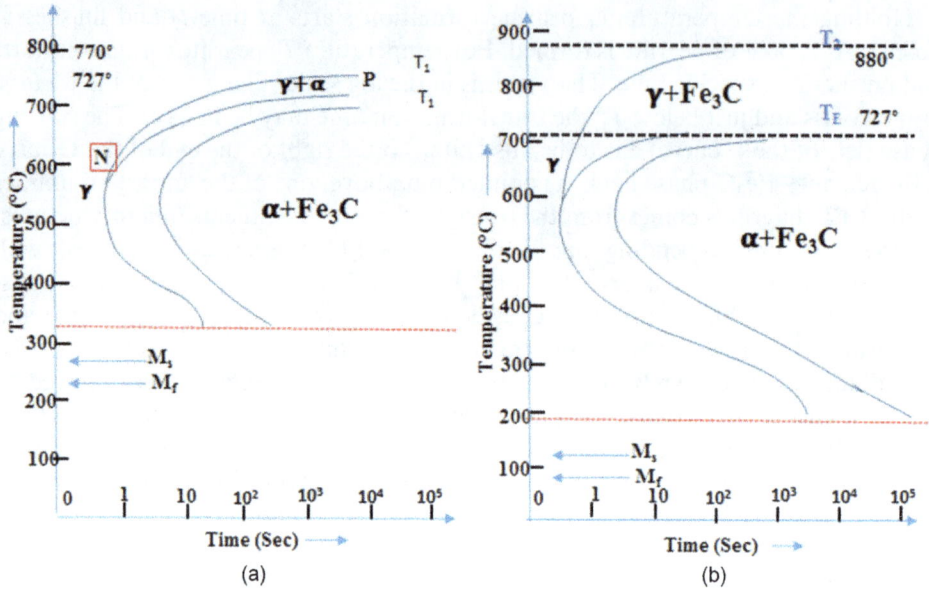

Figure 8.25. TTT diagrams for (a) hypo-eutectoid steel and (b) hypereutectoid steel.

cementite. As the transformation temperature decreases, the characteristic lamellar structure is maintained, but the spacing between the ferrite and carbide layers becomes increasingly small until the separate layers cannot be resolved with an optical microscope. As the temperature of transformation and the distance between the lamellae of the pearlite decrease, it is apparent that the hardness will increase.

Pearlite colonies are formed by nucleation and growth techniques. When the cooling curve reaches the pearlite starting temperature in the TTT curve, the pearlite starts to nucleate at the grain boundaries (see figure 8.26). Here the interlamellar spacing is the function of the temperature of transformation rather than time. The lower the quenching temperature is, the finer the pearlite colonies are, thus increasing the hardness.

8.3.4 Transformation to bainite

Between the nose region of approximately 510 °C and the M_s temperature, a new, dark-etching aggregate of ferrite and cementite appears. This structure, named after E C Bain, is called bainite. At the upper temperatures of the transformation range, it resembles pearlite and is known as upper or feathery bainite. At low temperatures it appears as a black needlelike structure resembling martensite and is known as lower or acicular bainite. The microstructure of the bainite is shown in figure 8.27.

Pearlite is nucleated by a carbide crystal and bainite is nucleated by a ferrite crystal, and this results in different growth patterns. Bainite is acicular in shape and is accompanied by surface distortions. At lower temperatures the carbide formed could be ε-carbide.

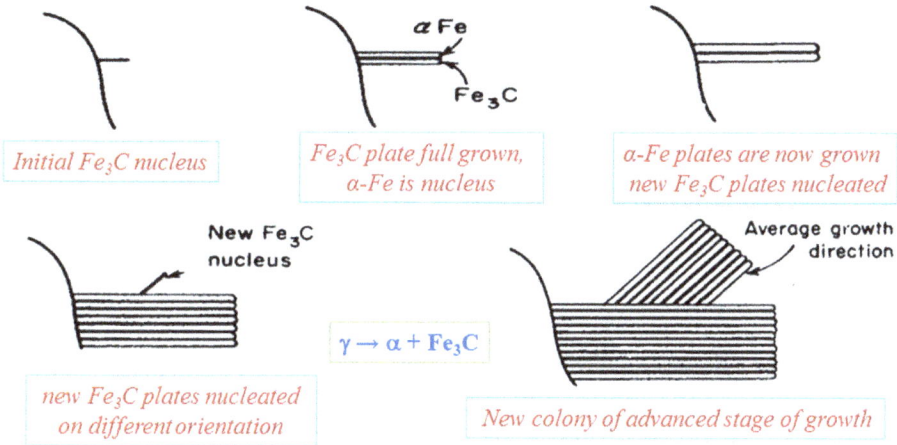

Figure 8.26. A schematic diagram of the growth of pearlite.

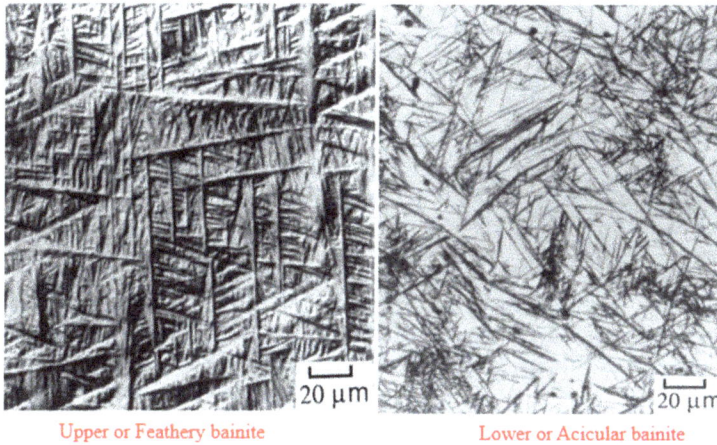

Upper or Feathery bainite   Lower or Acicular bainite

Figure 8.27. Microstructures of bainite. Reproduced with permission from Sajjadi and Zebarjad (2007)

8.3.5 Cooling curves and the *I–T* diagram

The following paragraphs refer to the curves in figure 8.28.

Curve 1. Cooling curve 1 shows a very slow cooling rate of conventional annealing. Transformation will start at x_1 and ends at x_1^*, and there is a slight temperature difference in the beginning and end transformation temperatures. There will be a slight difference in the fineness of the pearlite formed at the beginning and at the end.

Curve 2. Cooling curve 2 shows 'isothermal' cooling and was developed directly from the *I–T* diagram. The process is carried out by cooling the material rapidly from above the critical range to a predetermined temperature in the upper portion of the *I–T* diagram and holding for the time indicated to produce complete transformation. In contrast to conventional annealing, this treatment produces a more uniform microstructure and hardness.

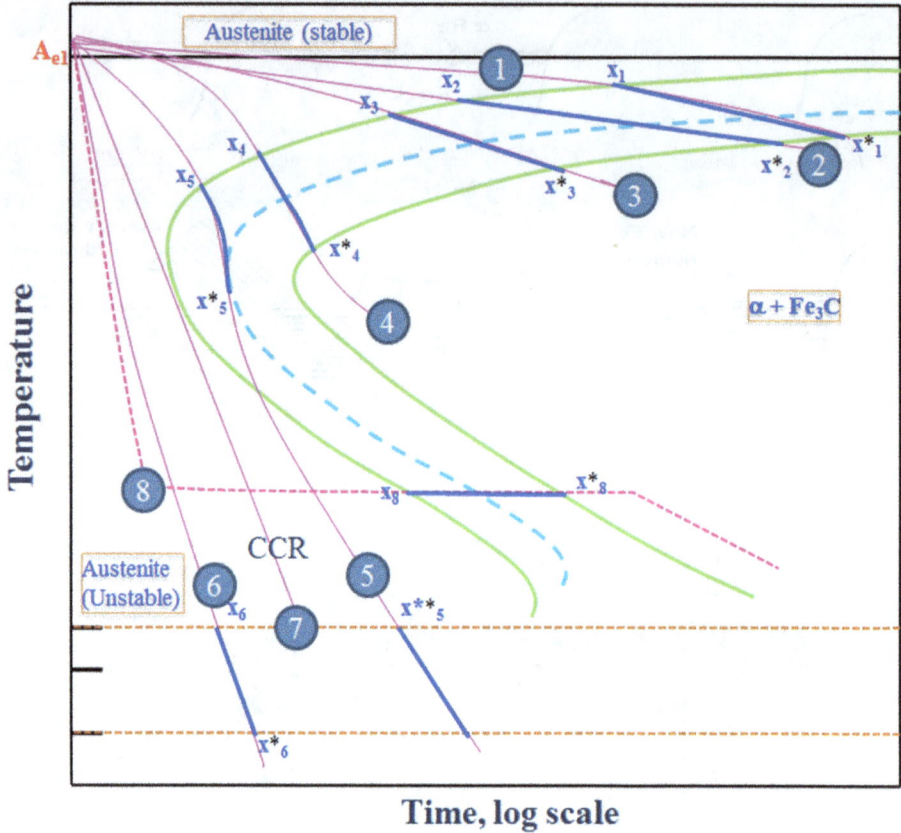

Figure 8.28. Different cooling rates (1–8) imposed on the TTT diagram for eutectoid steel. Reproduced with permission from Atlas of Isothermal Transformation Diagrams (United States Steel Corporation).

Curve 3. Cooling curve 3 has a faster cooling rate compared to annealing and may be considered typical of normalizing. The diagram indicates that transformation will start at x_3, with the formation of coarse pearlite, and ends at x_1^*, with the formation of medium pearlite. Here, we are clearly observing the temperature difference between x_3 and x_3^*, and x_1 and x_1^*, so the microstructure will show a greater variation in the fineness of pearlite.

Curve 4. Cooling curve 4, typical of a slow oil quench, is similar to curve 3, and the microstructure will be a mixture of medium and fine pearlite.

Curve 5. Cooling curve 5, typical of an intermediate cooling rate, will start to transform at x_5 to fine pearlite in a relatively short time. After a certain temperature, the cooling curve follows the direction of decreasing percent transformed. Since pearlite cannot form austenite on cooling, the transformation must stop at x_5^*. The microstructure at this point will consist of fine pearlite surrounded by austenitic grains. It will remain in this condition until the M_s line is crossed at x_5^{**}. The remaining austenite has now transformed into martensite. The final microstructure consists of martensite and fine nodular pearlite, largely concentrated along the original austenite grain boundaries.

Curve 6. Cooling curve 6, typical of drastic quench, is rapid enough to avoid transformation in the nose region. It remains austenitic until the M_s line is reached at x_6. Transformation to martensite will take place between the M_s and M_f lines. The final microstructure will be entirely martensite of high hardness.

Curve 7. Cooling curve 7, which is tangent to the nose, is the approximate CCR for steel. Any cooling rate slower than the one indicated will cut the curve above the nose and form some softer transformation product. Any cooling rate faster than the one illustrated will form only martensite. To obtain a fully martensitic structure, it is necessary to avoid transformation in the nose region.

Curve 8. It is possible to form 100% pearlite or 100% martensite by continuous cooling, but it is not possible to form 100% bainite. A complete bainitic structure may be formed only by cooling rapidly enough to miss the nose of the curve and then holding in the temperature range at which bainite is formed until transformation is complete. This is illustrated by cooling curve 8. It is apparent that continuously cooled steel samples will contain only small amounts of bainite, and this is probably the reason why this structure was not recognized until isothermal studies were conducted.

8.3.6 Factors affecting the TTT diagram

The main factors that affect the TTT diagram are the carbon percentage of the steel and the percentage of alloying elements. The grain size and heterogeneity of austenite also affects the TTT diagram to some extent.

Composition. With some limitations, an increase in carbon or alloy content, or an increase in the grain size of the austenite, always retards the transformation (moves curves to the right), at least at temperatures at or above the nose region. This in turn slows the CCR, making is easier to form martensite. This retardation is also reflected in the greater hardenability, or depth of penetration of hardness, of steel with higher alloy content or larger austenitic grain size.

Heterogeneity of austenite. Heterogeneous austenite increases the transformation time range, i.e. from the start to finish of the ferritic, pearlitic and bainitic ranges, and also increases the transformation temperature range in the case of martensitic transformation and bainitic transformation. Undissolved cementite and carbides act as powerful inoculants for pearlitic transformation. Therefore, heterogeneity in austenite increases the transformation time range in diffusional transformation and the temperature range of shear transformation products in the TTT diagram.

8.4 Continuous cooling transformation diagrams

The TTT diagrams are also called isothermal transformation diagrams, because the transformation times are representative of an isothermal hold treatment (following an instantaneous quench). However, in practical situations we follow heat treatments (T–t procedures/cycles) in which (typically) there are steps involving cooling of the sample. The cooling rate may or may not be constant. The rate of cooling may be slow (as in a furnace which has been switched off) or rapid (such as quenching in water).

Hence, in terms of practical utility, TTT curves have limitations and we need to draw separate diagrams called CCT diagrams, wherein the transformation times (also the products and microstructure) are noted using a constant rate cooling treatments. A diagram drawn for a given cooling rate (dT/dt) is typically used for a range of cooling rates (thus avoiding the need for a separate diagram for every cooling rate).

However, TTT diagrams are also often used for constant cooling rate experiments, keeping in mind the assumptions and approximations involved. An important difference between the CCT and TTT transformations is that in the CCT case bainite cannot form.

The CCT diagram for eutectoid steel is considered next.

8.4.1 Determination of the CCT diagram for eutectoid steel

CCT diagrams are determined by measuring some physical properties during continuous cooling, such as like specific volume and magnetic permeability. However, the majority of the work has been done through specific volume change by a dilatometric method. This method is supplemented by metallography and hardness measurements.

In dilatometry the test sample is austenitized in a specially designed furnace and then cooled in a controlled manner. Sample dilation is measured by a dial gauge/sensor. The slow cooling is controlled by furnace cooling, but a higher cooling rate can be controlled by gas quenching.

Cooling data are plotted as temperature versus time (see figure 8.29(a)). Dilation is recorded against temperature (see figure 8.29(b)). Any slope change indicates phase transformation. The fraction of transformation can be calculated roughly based on the dilation data, as explained below.

In general, the slope of the dilation curve remains unchanged while the amount of phase, or the relative amount of phases, in a phase mixture does not change during cooling (or heating). However, the sample can shrink or expand, i.e. dilation takes place purely due to thermal specific volume change because of a change in temperature.

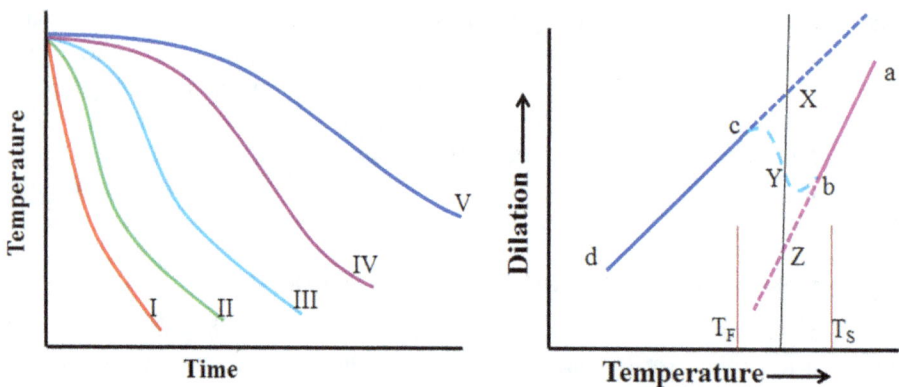

Figure 8.29. (a) Curves I–V indicate cooling curves at a higher cooling rate to lower cooling rate, respectively. (b) Dilation at different temperatures for a given cooling rate/schedule.

Therefore, in figure 8.29(b) dilation from *a* to *b* is due to the specific volume change of high temperature phase austenite. However, the slope of the curve changes at T_s. Therefore, transformation starts at T_s. Again, the slope of the curve from *c* to *d* is constant, but is different from the slope of the curve from *a* to *b*. This indicates that there is no phase transformation between the temperature from *c* to *d*, but the phase/phase mixture is different from the phase at *a* to *b*.

The slope of the dilation curve from *b* to *c* is variable with temperature. This indicates the change in the relative amount of phase due to cooling. The expansion is due to the formation of low-density phase(s). Some part of the dilation is compensated by purely thermal change due to cooling. Therefore, the dilation curve takes a complex shape, i.e. first the slope decreases and reaches a minimum value and then increases to the characteristic value of the phase mixture at *c*.

Therefore, phase transformation starts at *b*, i.e., at temperature T_s, and transformation ends or finishes at *c* or temperature T_f. The nature of the transformation has to be determined by metallography. When austenite fully transforms to a single product, then the amount of transformation is directly proportional to the relative change in length. For a mixture of products the percentage of austenite transformed may not be strictly proportional to the change in length, however, it is reasonable and is generally used.

For every type of transformation, the loci of the start points, isopercentage points and finish points give the transformation start line, isopercentage lines and finish line, respectively, and that results in the CCT diagram (see figure 8.29). Normally at the end of each cooling curve the hardness value of the resultant product is at room temperature. The types of phases obtained are shown in figure 8.30.

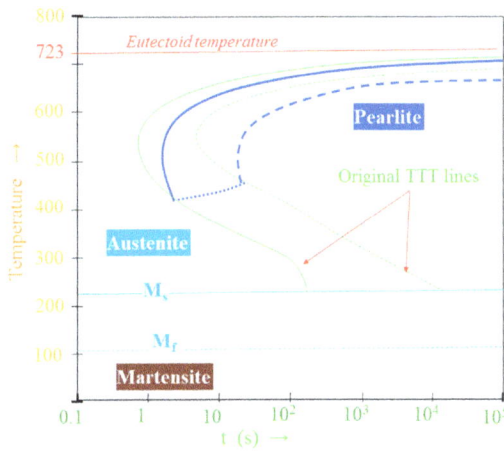

Figure 8.30. CCT diagram for eutectoid steel. Reproduced with permission from Atlas of Isothermal Transformation Diagrams (United States Steel Corporation).

8.4.2 Different cooling treatments for eutectoid steel

Figure 8.31 shows the different cooling operations generally performed on steels. Based on the cooling rate the product phase also varies. Typical cooling rate is given for different cooling processes.

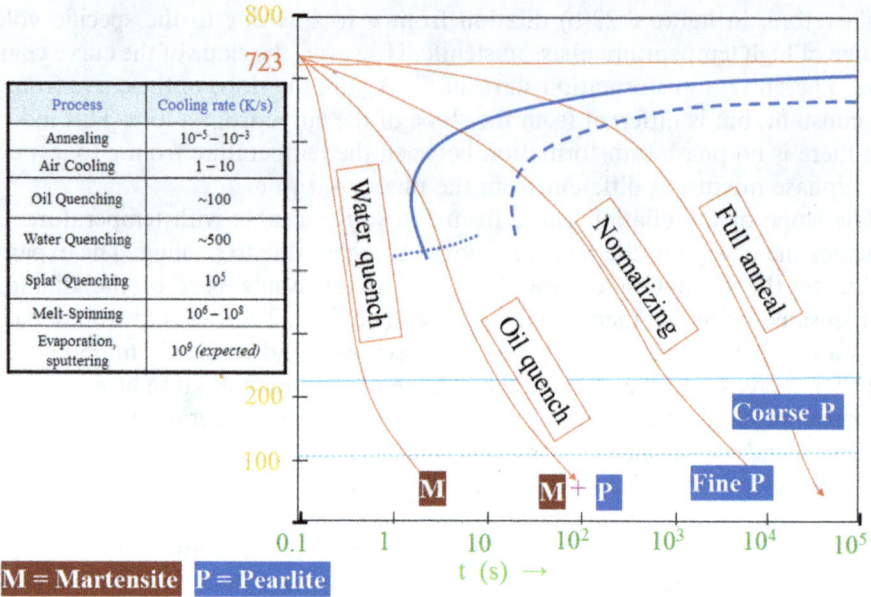

Process	Cooling rate (K/s)
Annealing	$10^{-5} - 10^{-3}$
Air Cooling	$1 - 10$
Oil Quenching	~100
Water Quenching	~500
Splat Quenching	10^5
Melt-Spinning	$10^6 - 10^8$
Evaporation, sputtering	10^9 *(expected)*

Figure 8.31. Different cooling rates applied on steel. Reproduced with permission of Professor Subramaniam, Indian Institute of Technology http://home.iitk.ac.in/~anandh/E-book/.

8.5 Quenching

8.5.1 Mechanism

The properties of materials are altered through different heat treating processes by varying the cooling rate followed by quenching. If the actual cooling rate (ACR) exceeds the CCR, only martensite will result. If the ACR is less than the CCR, the part will not completely harden. Thus the differences in cooling rates lead to variations in properties. At this point, it is necessary to understand the mechanism of heat removal during quenching.

A cooling curve shows the variation of temperature with time during quenching. A cooling rate, however, shows the rate of change of temperature with time. The cooling rate at any temperature may be obtained from the cooling curve by drawing a tangent to the curve at that temperature and determining the slope of the tangent. It is apparent that the cooling rate is constantly changing with time.

8.5.2 Stages of quenching

The different stages are illustrated in figure 8.32.

Vapor-blanket cooling stage. In this stage, the quenching medium is vaporized at the surface of the metal and forms a thin stable layer. Cooling in this stage occurs by conduction and radiation. As a vapor blanket exists around the material, the cooling rate is relatively slow.

Figure 8.32. Different stages of quenching: (a) vapor-blanket stage, (b) vapor transport cooling stage and (c) liquid cooling stage.

Vapor-transport cooling stage. This stage starts when the metal has cooled to a temperature at which the vapor film is no longer stable. The metal surface is wetted by the quenching medium and violent boiling of the quenching medium occurs around the metal. Heat is removed from the metal very rapidly as the latent heat of vaporization. This is the fastest stage of cooling.

Liquid cooling stage. This stage starts when the surface temperature of the metal reaches the boiling point of the quenching liquid. Vapor no longer forms, so cooling is by conduction and convection through the liquid. The rate of cooling is slowest at this stage.

8.5.3 Factors affecting quenching

Many factors determine the ACR. The most important are the type of quenching medium, the temperature of the quenching medium, the surface condition of the part, and the size and mass of the part.

Quenching medium
The cooling rate depends on the kind of quenching medium used during the hardening process. The commonly used quenching mediums are as follows:
  1. Brine solution (water solution of 10% sodium chloride).
  2. Tap water.
  3. Fused or liquid salts.
  4. Soluble oil and water solutions.
  5. Oil.
  6. Air.

The temperature of the quenching medium
Generally, as the temperature of the medium increases, the cooling rate decreases. This is due to the increase in the persistence of the vapor-blanket stage. This is particularly true for water and brine. In another case, an increase in the cooling rate with an increase in the temperature of the medium is observed. In the case of oil, as

the temperature of the oil increases, there is a tendency for the cooling rate to decrease due to the persistence of the vapor film. However, as the temperature of the oil increases, it also becomes more fluid, which increases the rate of heat conduction through the liquid.

Surface conditions

When the steel is exposed to an oxidizing atmosphere, because of the presence of water vapor or oxygen in the furnace, a layer of iron oxide called *scale* is formed. Experiments have shown that a thin layer of scale has very little effect on the ACR, but a thick layer of scale (0.005 inches deep) retards the ACR.

There is also the tendency for parts of the scale to peel off the surface when the piece is transferred from the furnace to the quench tank, thus giving rise to a variation in cooling rate at different points on the surface. The presence of scale only needs to be considered if the ACR is very close to the CCR.

To minimize the formation of scale, there are different methods depending on the part being heat treated, the type of furnace used, the availability of equipment and cost:

- Copper plating.
- Protective atmosphere.
- Liquid salt pots.
- Cast iron chips.

Size and mass

The ratio of surface area to mass is an important factor in determining the ACR, because only the surface of the metal stays in contact with the quenching medium. Thin plates and small diameter wires have a large ratio of surface area to mass and thus have rapid cooling rates:

$$\text{for a cylinder} \quad \frac{\text{surface area}}{\text{mass}} = \left(\frac{\pi D L}{(\pi/4) D^2 L \rho} \right)$$

The calculation shows that the ratio is inversely proportional to the diameter. If the diameter is increased, the ratio of surface area to mass decreases, and the cooling rate decreases. The heat in the interior of the piece must be removed by conduction through the body of the piece, eventually reaching the surface and the quenching medium. Therefore, the cooling rate in the interior is less than that at the surface.

If such a variation in cooling rates exists across the radius of a bar during cooling, it is to be anticipated that variations in hardness would be evident when the bars are cut and hardness surveys made on the cross section. So a considerable temperature difference exists between the surface and the center during quenching. Thus a hardness profile can be deduced during the quenching of a cylindrical specimen (see figure 8.33). This temperature difference will give rise to stresses during heat treatment called residual stresses, which may result in distortion and cracking of the piece.

8.5.4 The hardness profile of a cylinder from the surface to the interior

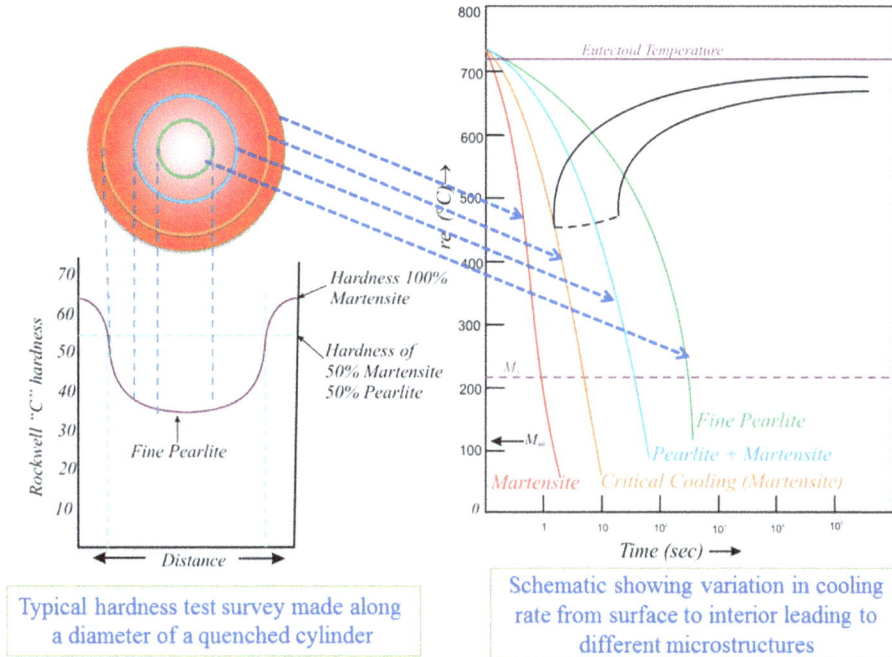

Typical hardness test survey made along a diameter of a quenched cylinder

Schematic showing variation in cooling rate from surface to interior leading to different microstructures

Figure 8.33. The difference in hardness resulting within a cylindrical bar due to differential cooling rates during quenching.

8.5.5 Hardenability

We have seen above that the hardness profile varies from the surface to the core. This leads to the important conclusion that varying the diameter of a cylinder with the same composition will give different hardness profiles, i.e. the depth of hardness is different for different sized samples. Generally, these diagrams are called hardness-penetration diagrams or hardness-traverse diagrams (see figure 8.34).

Hardenability is the ability of a steel to partially or completely transform from austenite to some fraction of martensite at a given depth below the surface, when cooled under a given condition. For example, a steel of a high hardenability can transform to a high fraction of martensite to depths of several millimeters under relatively slow cooling, such as an oil quench, whereas a steel of low hardenability may only form a high fraction of martensite to a depth of less than a millimeter, even under rapid cooling such as a water quench. Hardenability therefore describes the capacity of the steel to harden in depth under a given set of conditions.

The increase in the hardenability or depth of penetration of the hardness may be accomplished by either of two methods:

1. With the ACRs fixed, slow down the CCR (shift the I–T curve to the right) by adding alloying elements or coarsening the austenitic grain size.
2. With the I–T curve fixed, increase the ACRs by using a faster quenching medium or increasing circulation.

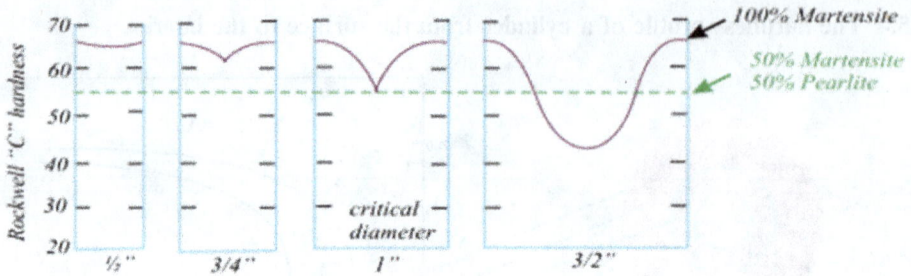

Figure 8.34. Hardness-penetration diagrams (or hardness-traverse diagrams) for steels with different diameters. Diameter is in increasing order from left to right. The amount and types of phases formed are also correlated with the hardness values. Reproduced with permission of Professor Subramaniam, Indian Institute of Technology http://home.iitk.ac.in/~anandh/E-book/.

Since increasing the cooling rates increases the danger of distortion or cracking, the addition of alloying elements is the more popular method of increasing hardenability.

Steels with high hardenability are needed for large high-strength components, such as large extruder screws for injection molding of polymers, pistons for rock breakers, aircraft undercarriage, etc. Steels with low hardenability may be used for smaller components, such as chisels and gears, etc. The most widely used method of determining the hardenability is the end-quench hardenability test or the Jominy test, commonly called the Jominy end-quench test.

8.5.6 The Jominy end-quench test

For conducting this test, a one inch round specimen four inches long is heated uniformly to the proper austenizing temperature. It is then removed from the furnace and placed in a fixture where a jet of water impinges on the bottom face of the sample.

After 10 min on the fixture, the specimen is removed and cut along the longitudinal direction. Then the Rockwell C scale hardness readings are taken at 1/16 inch intervals from the quenched end. The results are expressed as a curve of hardness values versus distance from the quenched end. A typical hardenability curve for eutectoid steel is shown in figure 8.35.

A number of Jominy end-quench samples are first end-quenched for a series of different times and then each of them (the whole sample) is quenched by complete immersion in water to freeze the already transformed structures.

Cooling curves are generated by putting a thermocouple at different locations and recording temperature against cooling time during end-quenching. The microstructures at the point where cooling curves are known are subsequently examined and measured using quantitative metallography. The hardness measurement is performed at each investigated point. Based on the metallographic information at the investigated point, the transformation start and finish temperatures and time are determined. The transformation temperature and time are also determined for a specific amount of transformation. These are located on cooling curves plotted in

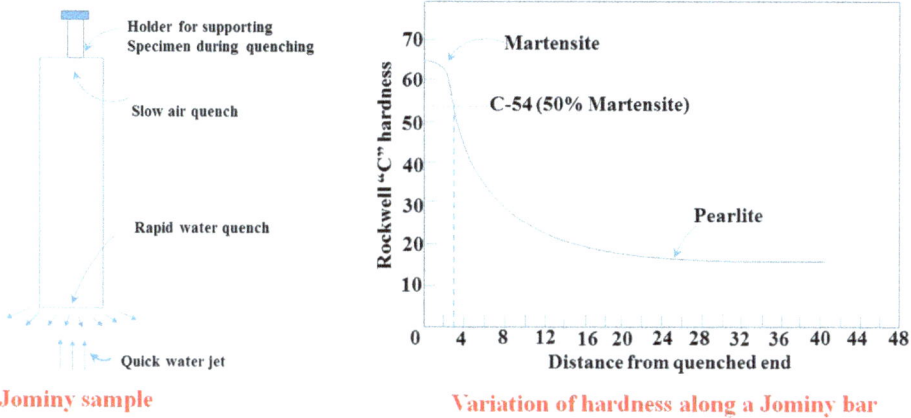

Jominy sample

Variation of hardness along a Jominy bar

Figure 8.35. The Jominy end-quench test and the variation of hardness along the Jominy bar.

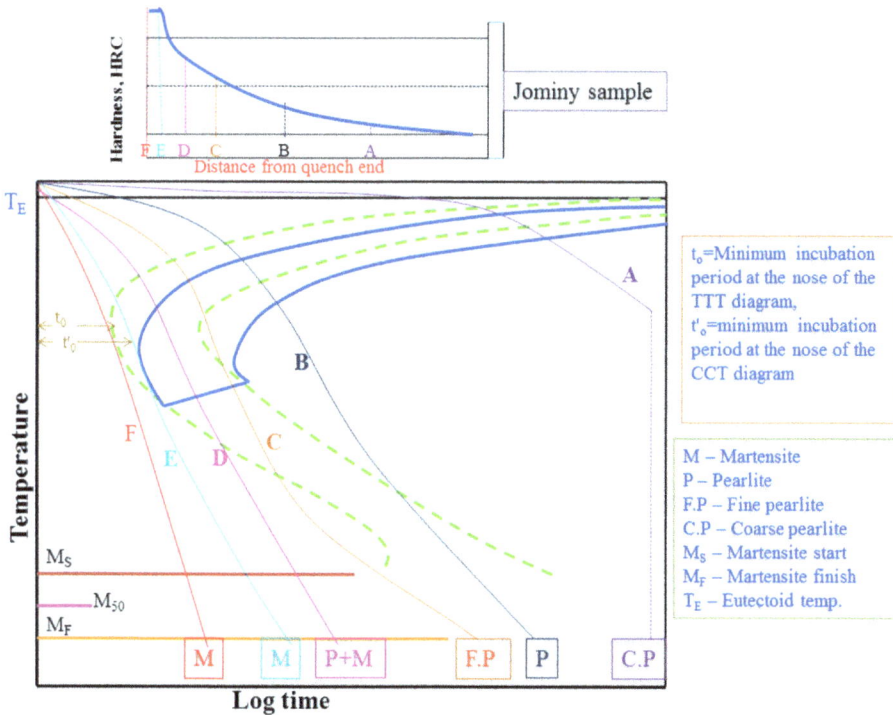

Figure 8.36. CCT diagram for different cooling rates in a Jominy end-quench test bar. Reproduced with permission of Dr Rampada Manna, Associate Professor, Indian Institute of Technology (Banaras Hindu University), Varanasi, India.

a temperature versus time diagram. The locus of the transformation start, finish or specific percentage of transformation generates the CCT diagram.

A, *B*, *C*, *D*, *E* and *F* are six different locations on the Jominy sample, shown in figure 8.36, which give six different cooling rates. The cooling rates *A*, *B*, *C*, *D*, *E* and *F* are in increasing order. The corresponding cooling curves are shown on the temperature log time plot. At the end of the cooling curve the phases are shown at room temperature. Variation in hardness with distance from the Jominy end is also shown in the figure.

8.6 Tempering

Tempering is a reheating treatment process to relieve the residual stresses and improve the ductility and toughness of steel. In other words, tempering is a process to optimize the strength and toughness of steel. With the increase in temperature, carbon diffusion becomes appreciable and the metastable martensite decomposes to more stable products. Four stages of tempering are known.

Stage 1. Stage 1 of temperature extends from room temperature to 200 °C. During this stage the martensite decomposes to two phases: a low-carbon martensite with 0.2% C is sometimes known as black martensite, and ε (hcp, $Fe_{2.4}C$), a transition carbide. With an increase in the carbon content of the steel, more ε-carbide forms. It precipitates in a very fine form and resolves only under an electron microscope. The hardening effect due to this precipitation is usually balanced by the softening effect associated with the loss of carbon in martensite.

Stage 2. Stage 2 tempering occurs in the range of 250–400 °C and changes the ε carbide to orthorhombic cementite, the low-carbon martensite becomes bcc ferrite and any retained austenite is transformed to lower bainite as a function of time. The carbides are too small to be resolved by an optical microscope, and the entire structure etches rapidly to a black mass formerly called troostite.

Stage 3. Stage 3 in tempering is in the range 400–650 °C. In this stage the growth of the cementite particles continues. This coalescence of the carbide particles allows more of the ferrite matrix to be seen, causing the sample to etch lighter than the low-temperature product. This structure, formerly known as sorbite, is resolvable at a magnification greater than 500×, under electron microscopy it is clearly visible.

Stage 4. Stage 4 tempering is in the range from 650–720 °C. This produces large, globular cementite particles. This structure is very soft and tough, and is similar to the spheroidized cementite structure. Spheroidite has been the softest yet toughest structure that steel can have.

For many years, metallurgists divided the tempering processes into definite stages. The microstructures appearing on these stages was given names such as black martensite, troostite and sorbite. However, the changes in microstructures are so gradual that it is more realistic to call the product of tempering at any temperature simply tempered martensite. The transformation products from austenite by different heat treatments are given in figure 8.37.

Tempering of some steels may result in a reduction of toughness, which is known as temper embrittlement. This may be avoided by (1) compositional control and/or (2) tempering above 575 °C or below 375 °C, followed by quenching to room temperature. The effect is greatest in martensite structures, less severe in bainite structures and least severe in pearlite structures. It appears to be associated with the segregation of solute atoms to the grain boundaries lowering the boundary strength.

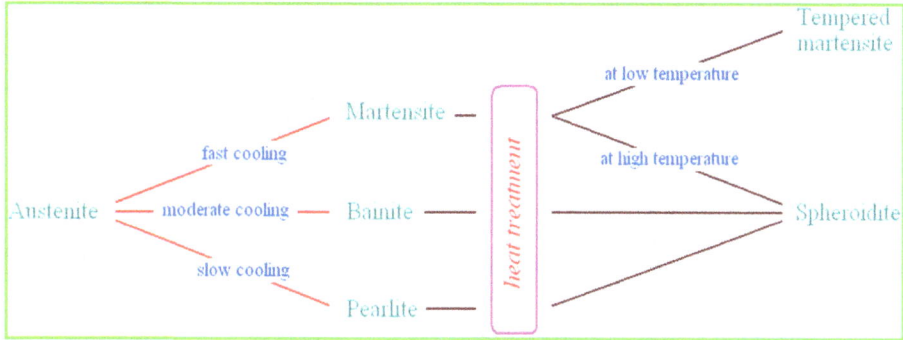

Figure 8.37. Transformation of austenite into various products by different heat treatments.

Impurities responsible for temper brittleness are P, Sn, Sb and As. Si reduces the risk of embrittlement by carbide formation. Mo has a stabilizing effect on carbides and is also used to minimize the risk of temper brittleness in low-alloy steels:

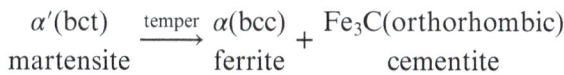

$$\underset{\text{martensite}}{\alpha'(\text{bct})} \xrightarrow{\text{temper}} \underset{\text{ferrite}}{\alpha(\text{bcc})} + \underset{\text{cementite}}{\text{Fe}_3\text{C}(\text{orthorhombic})}$$

8.6.1 Martempering

This heat treatment is given to oil hardenable and air hardenable steels, and thin sections of water hardenable steel samples to produce martensite with minimal differential thermal and transformation stress to avoid distortion and cracking. The steel should have a reasonable incubation period at the nose of its TTT diagram and a long bainitic bay. The sample is quenched above the M_s temperature in a salt bath to reduce thermal stress (instead of cooling below M_f directly). The surface cooling rate is greater than at the center. The cooling schedule is such that the cooling curves pass behind without touching the nose of the TTT diagram. The sample is isothermally held at the bainitic bay such that the differential cooling rate at the center and the surface become equalized after some time. The sample is allowed to cool by air through M_s–M_f such that martensite forms both at the surface and center at the same time due to a smaller temperature difference, and thereby avoids transformation stress due to volume expansion. The sample is given tempering treatment at a suitable temperature. Figure 8.38 explains this process.

8.6.2 Austempering

Austempering heat treatment is applied to steels to produce lower bainite concentrations in high-carbon steel without any distortion or cracking to the sample. The heat treatment involves the cooling of austenite rapidly in a bath maintained at a lower bainitic temperature (above M_s; avoiding the nose of the TTT diagram) and holding it there to equalize the surface and center temperature up to the bainitic finish time.

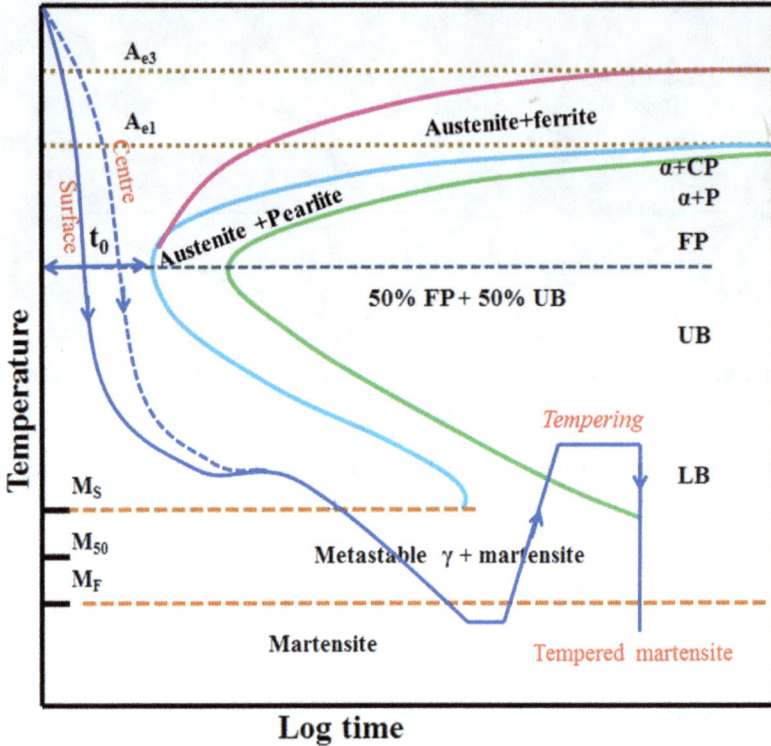

Figure 8.38. Martempering operation. Note that at one point the cooling rates for the surface and center become equal. Reproduced with permission of Dr Rampada Manna, Associate Professor, Indian Institute of Technology (Banaras Hindu University), Varanasi, India.

At the end of the bainitic transformation the sample is air cooled. The microstructure contains fully lower bainite. This heat treatment is given to steels containing 0.5–1.2 wt% C and low-alloy steels. The product hardness and strength are comparable to hardened and tempered martensite with improved ductility and toughness, and uniform mechanical properties. Products do not required tempering. Figure 8.39 shows the austempering process.

8.7 The role of alloying elements

Alloying elements are added to plain carbon steels in order to enhance the properties of the steels. Alloying elements cause segregation or phase separation in the steel. By creating solid solutions the alloying elements strengthen the steels.

8.7.1 Sample elements and their role

- Phosphorus (P) dissolves in ferrite to form a brittle phase named iron phosphide. Iron phosphide causes cold shortness in the steels.
- Sulfur (S) forms iron sulfide at the grain boundaries of ferrite and pearlite. This lowers the ductility during hot forging, i.e. hot shortness.

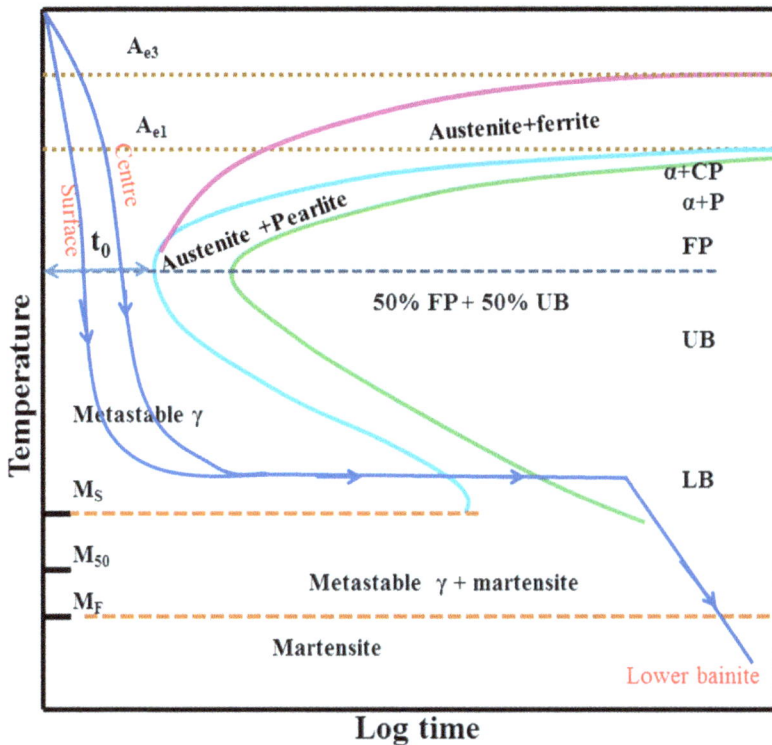

Figure 8.39. Austempering process. A holding temperature is shown where the cooling rates for both the surface and the center equalize. Reproduced with permission of Dr Rampada Manna, Associate Professor, Indian Institute of Technology (Banaras Hindu University), Varanasi, India.

- Silicon (Si) percentages between 0.2% and 0.4% improve the hardness, elastic modulus and ultimate tensile strength (UTS).
- Copper (Cu) has a solubility of 0.8% in ferrite. This causes precipitation hardening in steels.
- Lead (Pb) is insoluble in steel.
- Chromium (Cr) in steels is a ferrite stabilizer. It forms carbides, therefore increasing the strength and hardness. An addition of more than 11% forms passive films which give corrosion resistance to the steel.
- Nickle (Ni) is an austenite stabilizer in steels. Strength and toughness in steels are improved by the addition of nickel as an alloying element.
- Molybdenum (Mo) can be dissolved in both austenite and ferrite. It is a carbide former, so increases the strength and hardenability of the steels. Tempering embrittlement is lowered by the addition of molybdenum as an alloying element and the high temperature strength is also increased by adding Mo.

Some alloying elements can be classified as ferrite stabilizers and austenite stabilizers. An example of a ferrite stabilizer is chromium. With an increase in the percentage of chromium the area of the austenite region is squeezed. Manganese is

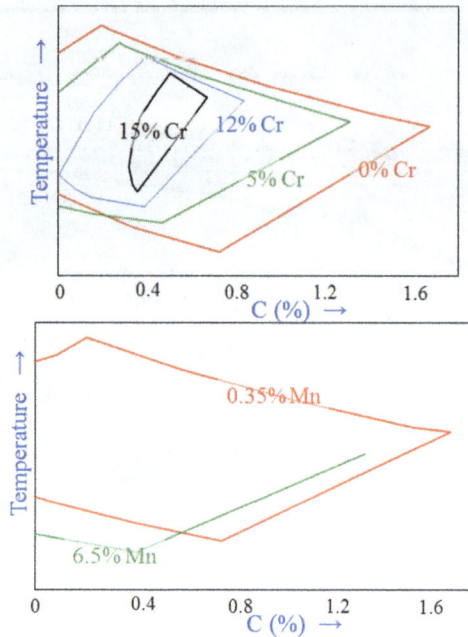

Figure 8.40. The effect of Cr (left) and Mn (right) on steels. It is clear that chromium is a ferritic stabilizer, whereas manganese is an austenitic stabilizer. Reproduced from American Society for Metals (1948) with permission.

an austenite stabilizer. The presence of manganese as an alloying element increases the austenitic region in the phase diagram. These trends are illustrated in figure 8.40.

8.8 Surface-hardening treatments

Numerous industrial applications require a hard-wearing resistant surface called the *case*, and a relatively soft, tougher part inside called the *core* (an example of this is gears). They are caused by two different processes. These are thermochemical and thermomechanical treatments. Thermochemical treatment is related to a change in chemical composition. In thermomechanical treatments there is no change of chemical composition of the steel and it is essentially a shallow-hardening method. This topic is outside the scope of this book, the reader may refer to the further reading section.

8.9 The iron–graphite phase diagram

8.9.1 History

Cast iron has its earliest origins in China between 700 and 800 BCE. Until this period ancient furnaces could not reach sufficiently high temperatures. The use of this newly discovered form of iron varied from simple tools to a complex chain suspension bridge erected in approximately 56 CE.

The next significant development in cast iron was the first use of coke in 1730 by an English founder called Abraham Darby. Due to this revolution, better casts were available for more versatile roles, such as James Watt's first steam engine, constructed in 1794.

In 1810, Swedish chemist Jöns Jacob Berzelius and German physicist Friedrich Stromeyer discovered that by adding silicon to the furnace, along with scrap and pig iron, consistently stronger cast iron is produced. In 1885 Thomas Turner added ferrosilicon to white iron to produce stronger gray iron castings. In the late twentieth century the major uses of cast irons consisted of pipes, thermal containment units, and certain machine or building entities which needed to absorb continuous vibrations.

8.9.2 Introduction

The true equilibrium diagram for iron and carbon is generally considered as an iron–graphite phase diagram. Earlier we learned about the iron–iron carbide phase diagram, which is not a true equilibrium diagram, generally it is called the metastable iron–iron carbide phase diagram. Cementite (Fe_3C) is a metastable compound, and under some circumstances, it can be made to dissociate or decompose to form ferrite and graphite, according to the reaction

$$Fe_3C \rightarrow 3Fe + C.$$

To explain cast irons we will refer to both the iron–iron carbide phase diagram and the iron–graphite phase diagram (see figure 8.41). Cast irons are a class of ferrous alloys with carbon contents above 2.14 wt%; in practice, however, most cast irons contain between 3.0 and 4.5 wt% C and, in addition, other alloying elements. The ductility of cast irons is very low and it is brittle, it cannot be rolled, drawn or worked at room temperature. However, cast irons melt readily and can be cast into

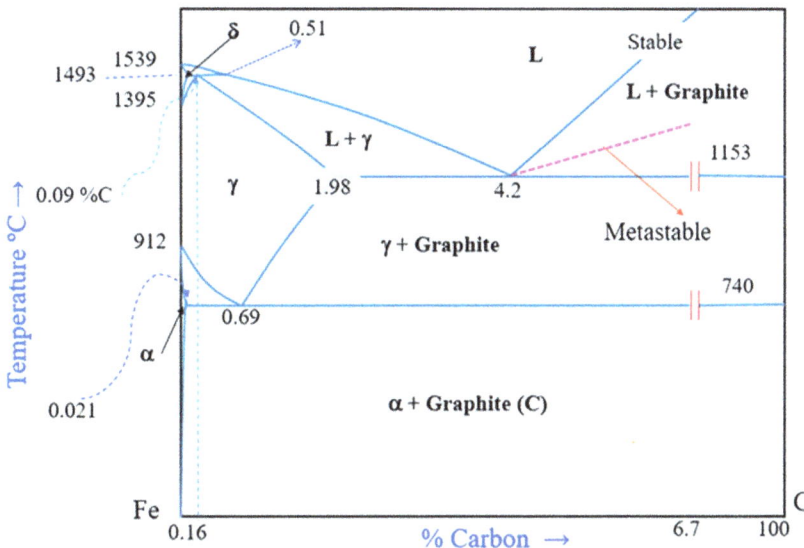

Figure 8.41. The iron–graphite phase diagram.

complicated shapes, which are usually machined to final dimensions. Since casting is the only suitable process applicable to these alloys, they are known as cast irons.

8.9.3 Classification of cast irons

Figure 8.42 shows the classification of cast irons.

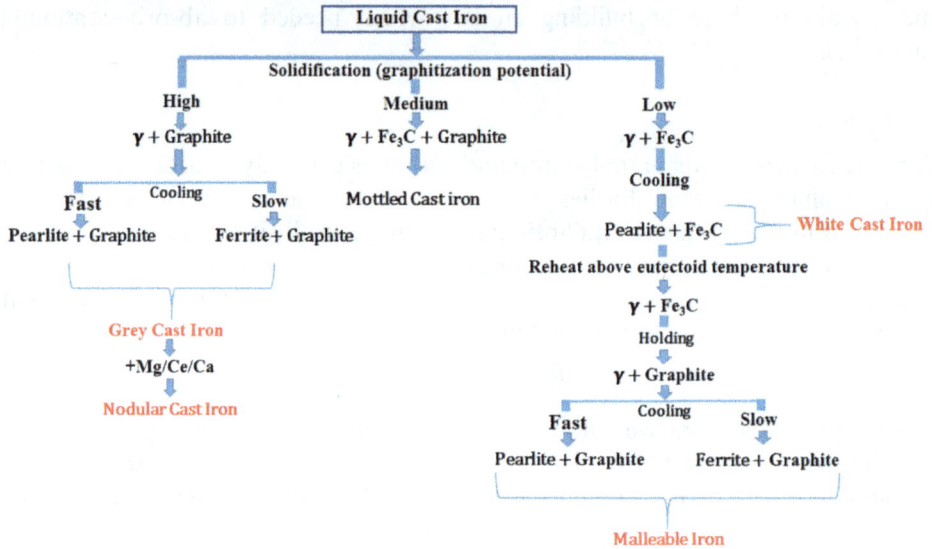

Figure 8.42. Classification of cast irons.

White cast iron

In white cast iron all the C is in the combined form as Fe_3C (cementite). The typical microstructure of white cast iron (figure 8.43), consisting of dendrites of transformed austenite (pearlite) in a white inter-dendritic network of cementite. White cast iron contains a relatively large amount of cementite as a continuous inter-dendritic network, it makes the cast iron hard and wear-resistant but extremely brittle and difficult to machine. 'Completely white' cast irons are limited in engineering applications because of this brittleness and lack of machinability. White cast iron is mainly used in liners for cement mixers, ball mills and extrusion nozzles. A large tonnage of white cast iron is used as a starting material for the manufacture of malleable cast iron. Hardness: 375–600 BHN. Tensile strength: 135–480 MPa. Compressive strength: 1380–1725 MPa.

Gray cast iron

This group is one of the most widely used alloys of iron, containing between 2.5% and 4% carbon with other alloying elements. In the manufacture of gray cast iron the tendency of cementite to separate into graphite and austenite or ferrite is favored by controlling alloy additions and cooling rates. These alloys solidify by first forming primary austenite. The initial appearance of combined carbon is in the

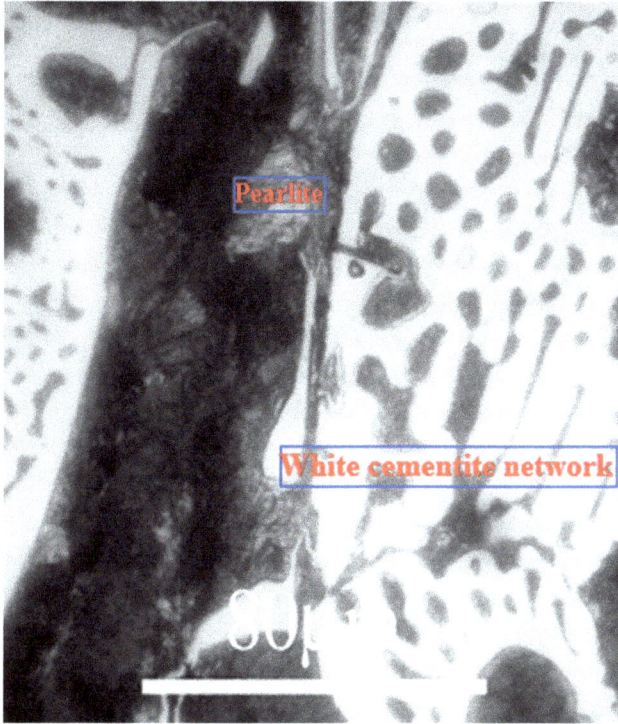

Figure 8.43. Microstructure of white cast iron. The white network formed as cementite and the darker phase corresponds to pearlite. Copyright DoITPoMS Micrograph Library, University of Cambridge. Reproduced with permission.

cementite resulting from the eutectic reaction. With proper control of carbon content, temperature and the proper amount of graphitizing elements, notably silicon, magnesium and cerium, the alloy will follow the stable iron–graphite equilibrium diagram. For most of these cast irons, the graphite exists in the form of flakes (similar to corn flakes), which are normally surrounded by an α-ferrite or pearlite matrix. Mechanically, gray iron is comparatively weak and brittle in tension as a consequence of its microstructure; the tips of the graphite flakes are sharp and pointed, and may serve as points of stress concentration when an external tensile stress is applied. Its strength and ductility are much higher under compressive loads. Gray irons are very effective in damping vibrational energy. The base structures for machines and heavy equipment that are exposed to vibrations are frequently constructed of this material. In addition, gray irons exhibit a high resistance to wear. Furthermore, in the molten state they have a high fluidity at casting temperature, which permits the casting of pieces with intricate shapes; also, casting shrinkage is low. Finally, and perhaps most importantly, gray cast irons are among the least expensive of all metallic materials. Gray irons have different types of microstructures, which may be generated by the adjustment of composition and/ or by using an appropriate heat treatment. For example, lowering the silicon content or increasing the cooling rate may prevent the complete dissociation of

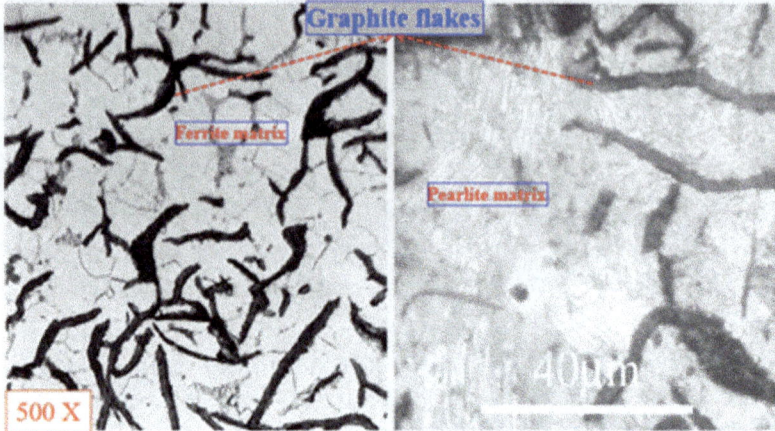

Figure 8.44. Microstructure of grey cast iron. Graphite flakes are observable in the matrix of the white ferrite network. Copyright DoITPoMS Micrograph Library, University of Cambridge. Reproduced with permission.

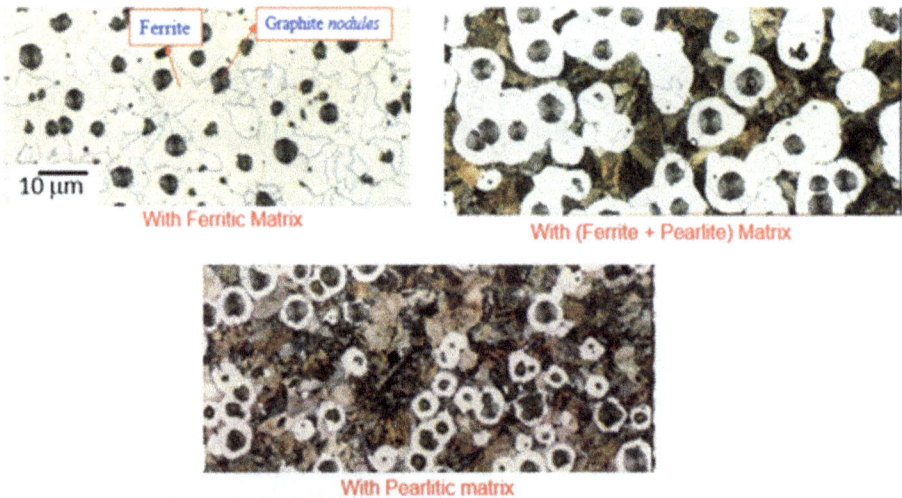

Figure 8.45. Nodular cast iron microstructure containing graphite nodules in a ferrite matrix, with a ferrite and perlite matrix, and in an only pearlitic matrix are shown. Reproduced with permission of Professor Subramaniam, Indian Institute of Technology http://home.iitk.ac.in/~anandh/E-book/.

cementite to form graphite. Under these circumstances the microstructure consists of graphite flakes embedded in a pearlite matrix. The microstructure of gray cast iron is shown in figure 8.44.

Ductile cast iron
Ductile cast iron is also called nodular cast iron, spheroidal graphite cast iron and spherulitic cast iron. Graphite nodules are present instead of flakes (see figure 8.45). Mg, Ce or Ca (or other spheroidizing) elements are added to nodularize the graphite

present. The elements added to promote spheroidization react with the solute in the liquid to form heterogeneous nucleation sites. The alloying elements are injected into the mould before pouring. It is thought that by the modification of the interfacial energy the 'c' and 'a' growth directions are made comparable, leading to spheroidal graphite morphology. The graphite phase usually nucleates in the liquid pocket created by the pro-eutectic γ.

Malleable cast iron

As we discussed in earlier chapters, the cementite is actually a metastable phase. There is a tendency for cementite to decompose into iron and carbon. This tendency to form free carbon is the basis for the manufacture of malleable cast iron. The reaction $Fe_3C \rightarrow 3Fe + C$ is favored by elevated temperatures, the existence of solid

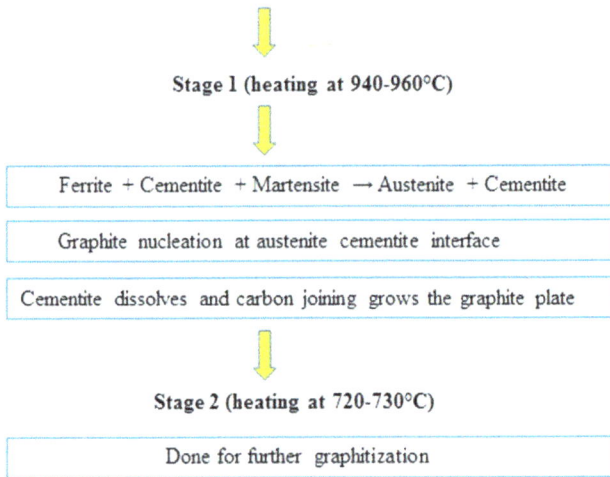

Stages involved in the preparation of malleable cast iron

Stage 1 (heating at 940-960°C)

Ferrite + Cementite + Martensite ⟶ Austenite + Cementite

Graphite nucleation at austenite cementite interface

Cementite dissolves and carbon joining grows the graphite plate

Stage 2 (heating at 720-730°C)

Done for further graphitization

(a)

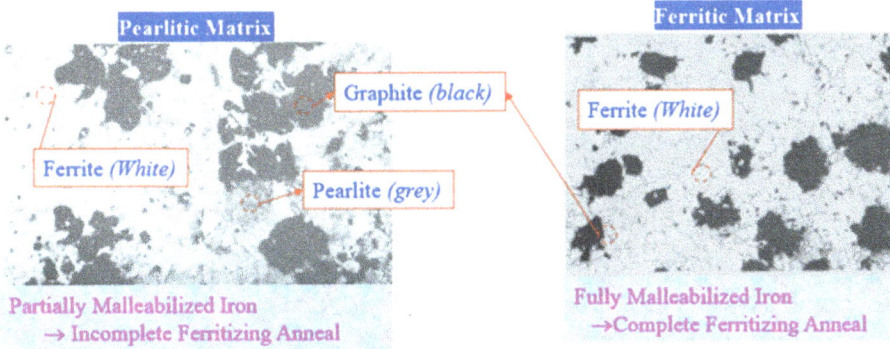

(b)

Figure 8.46. (a) The stages involved in the preparation of malleable cast iron and (b) the microstructures of malleable cast iron with a pearlitic matrix. Reproduced with permission of Professor Subramaniam, Indian Institute of Technology http://home.iitk.ac.in/~anandh/E-book/.

non-metallic impurities, higher carbon contents and the presence of elements that aid the decomposition of Fe_3C (see figure 8.46(a)). The microstructure for malleable cast iron is given in figure 8.46(b).

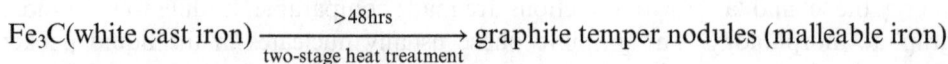

$$Fe_3C(\text{white cast iron}) \xrightarrow[\text{two-stage heat treatment}]{>48\text{hrs}} \text{graphite temper nodules (malleable iron)}$$

Compacted graphite cast iron

A relatively recent addition to the family of cast irons is compacted graphite iron (CGI). Microstructurally, the graphite in CGI alloys has a worm-like (or vermicular) shape; a typical CGI microstructure is shown in figure 8.47. In a sense, this microstructure is intermediate between that of gray iron and ductile iron and, in fact, some of the graphite (less than 20%) may be present as nodules. The chemistries of CGIs are more complex than for the other cast iron types; the compositions of Mg, Ce and other additives must be controlled so as to produce a microstructure that consists of the worm-like graphite particles, while at the same time limiting the degree of graphite nodularity and preventing the formation of graphite flakes. Furthermore, depending on the heat treatment, the matrix phase will be pearlite and/or ferrite. CGIs are now being used in a number of important applications, these include: diesel engine blocks, exhaust manifolds, gearbox housings, brake discs for high speed trains and flywheels.

Chilled cast iron

Chilled iron castings are made by casting the molten metal against a metal chiller, resulting in a surface of white cast iron. This hard, abrasion-resistant white iron surface or case is backed up by a softer gray iron core. This case–core structure is obtained by careful control of the overall alloy composition and adjustment of the cooling rate.

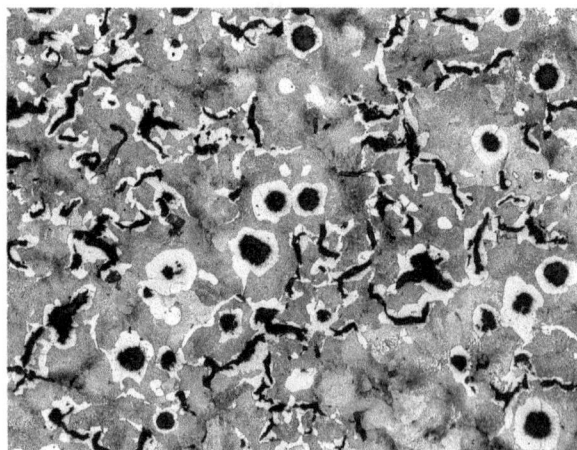

Figure 8.47. CGI microstructure.

Mottled cast iron
Mottled cast iron solidifies at a rate with extremes between those for chilled and gray irons, thus exhibiting microstructural and metallurgical characteristics of both.

Alloy cast iron
An alloy cast iron is one which contains a specially added element or elements in sufficient quantities to produce a measurable modification in the physical or mechanical properties. Elements normally obtained from raw materials, such as silicon, manganese, sulfur and phosphorus, are not considered alloy additions. Alloying elements are added to cast iron for special properties such as resistance to corrosion, heat or wear, and to improve the mechanical properties. Most alloying elements in cast iron will accelerate or retard graphitization, and this is one of the important reasons for alloying. The most common alloying elements are Cr, Cu, Mo, Ni and V.

High-alloy graphitic irons
High-alloy graphitic irons are produced with microstructures consisting of both flake and nodule structures. These are mainly utilized for applications requiring a combination of high strength and corrosion resistance.

Further reading

American Iron and Steel Institute 1970 *Steel Products Manual. Alloy Steel: Semifinished, Hot Rolled and Cold Finished Bars* (New York: American Iron and Steel Institute)
American Society for Metals 1948 *Metals Handbook* (Metals Park, OH: American Society for Metals)
ASM International 1990 *Properties and Selection: Iron, Steels and High Performance Alloys* (*Metals Handbook* vol 1) 10th edn (Materials Park, OH: ASM International)
Avner S H 1997 *Introduction to Physical Metallurgy* (New York: McGraw-Hill)
Callister W D 2007 *Callister's Materials Science and Engineering* (Indian Adaptation adapted by R Balasubramaniam) (New Delhi: Wiley)
Grossmann M A 1952 *Elements of Hardenability* (Metals Park, OH: American Society for Metals)
Guy A G 1959 *Elements of Physical Metallurgy* 2nd edn (Reading, MA: Addison-Wesley)
Hultgren R 1952 *Fundamentals of Physical Metallurgy* (Englewood Cliffs, NJ: Prentice-Hall)
Hume-Rothery W 1966 *Structure of Alloys and Iron* (New York: Pergamon)
Kingery W D, Bowen H K and Uhlmann D R 1976 *Introduction to Ceramics* (New York: Wiley) chs 8–10
Manna R TTT diagrams (Lecture notes) Cambridge University, UK, www.msm.cam.ac.uk
McCrum N G, Buckley C P and Bucknall C B 1988 *Principles of Polymer Engineering* (Oxford: Oxford University Press)
Porter D A, Eastererling K E and Sherif M Y 2009 *Phase Transformations in Metals and Alloys* (Boca Raton, FL: CRC Press Taylor and Francis Group)
Raghavan V 1987 *Solid State Phase Transformations* (New Delhi: Prentice Hall of India)
Raghavan V 2004 *Materials Science and Engineering* 5th edn (Englewood Cliffs, NJ: Prentice-Hall)
Reed-Hill R E 1964 *Physical Metallurgy Principles* (New York: Van Nostrand Reinhold)
Roberts-Austen W C 1897 Report 4 *Proc. Inst. Mech. Eng.* 33–100

Roozeboom H W B 1900 *Metallograph.* **3** 293–300

Sajjadi S A and Zebarjad S M 2007 Isothermal transformation of austenite to bainite in high carbon steels *J. Mater. Proc. Technol.* **189** 107–13

Sharma R C 1996 *Principles of Heat Treatment of Steels* (New Delhi: New Age International (P) LTD)

Singh V 2009 *Heat Treatment of Metals* (New Delhi: Standard Publishers Distributors)

Chapter 9

Physical metallurgy of non-ferrous alloys

Metallic materials, when considered in a broad sense, may be divided into two main groups, ferrous and non-ferrous. The ferrous materials are iron-based and non-ferrous materials have some elements other than iron as the principle constituent. The bulk of non-ferrous materials is made up of the alloys of copper, aluminum, titanium, nickel and other non-ferrous metals and alloys that are used to a lesser extent. This chapter will focus on the more important non-ferrous metals and alloys.

9.1 Copper and its alloys

9.1.1 Introduction

The most important properties of copper are its high electrical and thermal conductivity, good corrosion resistance, machinability, strength and easy fabrication. Most of the copper used for electrical conductors contains over 99.9% copper and is identified as electrolytic tough-pitch (ETP) copper or oxygen-free high-conductivity copper. ETP copper contains from 0.02% to 0.05% oxygen, which is combined with copper as the compound cuprous oxide (Cu_2O). As cast, copper oxide and copper form an inter-dendritic mixture. After working and annealing, the inter-dendritic network is destroyed and the strength is improved. Oxygen-free copper is used in electronic tubes or similar applications because it makes a perfect seal to glass. The alloys of copper covered in this chapter are shown in figure 9.1.

Arsenical copper: 0.3% arsenic has improved resistance to specific corrosive conditions.

Free cutting copper: 0.6% tellurium provides excellent machining properties.

Silver-bearing copper: 7–30 oz ton^{-1}. Silver raises the recrystallization temperature of copper, thus preventing softening during soldering of commutators.

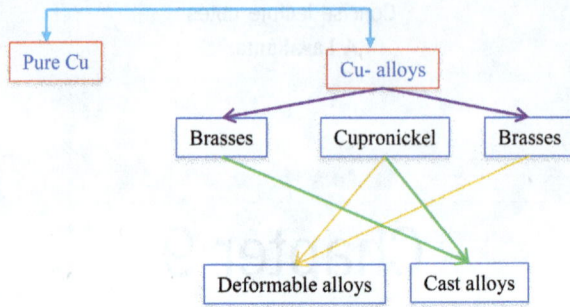

Figure 9.1. Classification of Cu and Cu alloys.

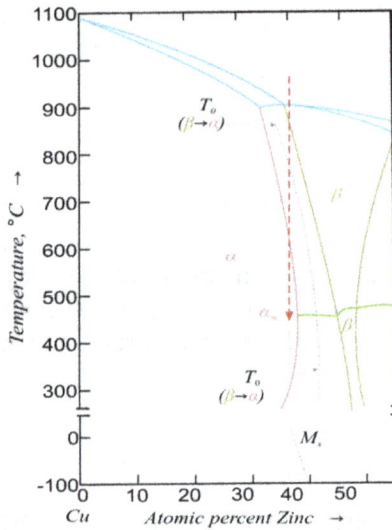

Figure 9.2. Cu-rich portion of the Cu–Zn phase diagram used in commercial applications. Reproduced from American Society for Metals (1948) with permission.

9.1.2 Brasses

Brasses are essentially alloys of copper and zinc. Some of these alloys have small amounts of other elements such as lead, tin or aluminum. The portion of the binary copper zinc phase diagram which is applicable to commercial alloys is shown in figure 9.2. The solubility of zinc in the alpha (α) solid solution increases from 32.5% at 910 °C to about 39% at 455 °C. Since copper is fcc, the α solid solution is fcc, and the beta (β) phase is a bcc electron compound and undergoes ordering, indicated by a dot-dashed line, in the region of 455–470 °C. On cooling in this temperature range the bcc β phase, with copper and zinc atoms randomly dispersed at lattice points, changes continuously to the ordered structure β', which is still bcc.

Alpha brasses
Alpha brasses have the copper atoms at the corners and zinc atoms at the centers of the unit cubes. The ordering reaction is so rapid that it cannot be retarded or

Alpha Brasses
(up to 36% Zn)

Yellow Brasses

Red Brasses

1. These contain 20-36 % Zn
2. It is common practice to stress relief anneal these brasses after severe cold working to prevent season cracking.
3. Season cracking or stress corrosion cracking is due to the high residual stresses left in the brass as a result of cold working.
4. Dezincification and Plug-type dezincification
5. Most widely used yellow α brasses are cartridge brass (70Cu-30Zn) and yellow brass (65Cu-35Zn)

1. These contain between 5 and 20 percent zinc.
2. The most common low zinc brasses are gliding metal (95Cu-5Zn), commercial , commercial bronze (90Cu-10Zn), red brass (85Cu-15Zn) and low brass (80Cu-20Zn)

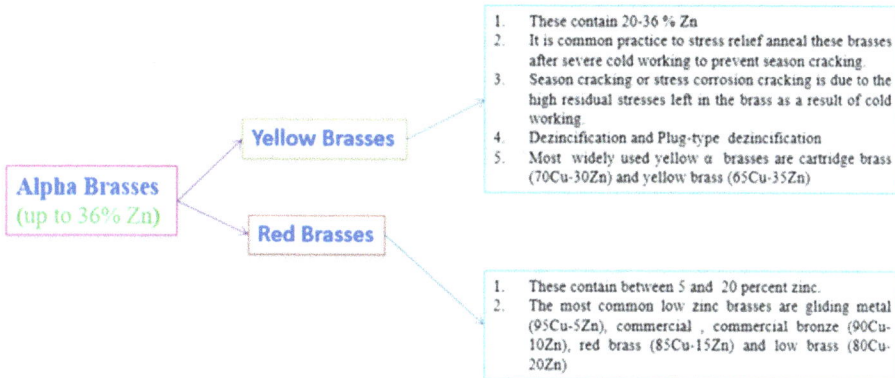

Figure 9.3. Classification of brasses.

prevented by quenching. The best combination of strength and ductility is obtained in 70Cu–30Zn brass. The commercial brasses can be divided into two groups, brasses for cold working (α brasses) and brasses for hot working (α plus β brasses). Alpha brasses can further be divided into other types. This classification is given in figure 9.3.

Alpha + beta brasses contain from 54% to 62% copper and this alloy consists of two phases, α and β'. The β' phase is harder and more brittle at room temperature than α; therefore, these alloys are more difficult to cold work than the α brasses. At elevated temperatures the β phase becomes very plastic, and since most of these alloys may be heated into the single-phase β region, they have excellent hot-working properties. The most widely used α + β' brass is Muntz metal (60Cu–40Zn), which has high strength and excellent hot-working properties.

The main types of alpha brasses are as follows:
- Muntz metal is used in condenser heads, perforated metal and architectural work. Free-cutting brass (61.5Cu–35.5Zn–3Pb) has the best machinability of any brass combined with good mechanical and corrosion resistant properties. The microstructure of Muntz metal is given in figure 9.4.
- Forging brass (60Cu–38Zn–2Pb) has the best hot-working properties of any brass and is used for hot forgings, hardware and plumbing parts.
- Architectural bronze (57Cu–40Zn–3Pb) has excellent forging and free matching properties.
- Naval brass (60Cu–39.25Zn–0.75Sn), also known as tobin bronze, has increased resistance to salt water corrosion and is used for condenser plates, welding rod, propeller shafts, piston rods and valve systems.
- Manganese bronze (58.5Cu–39Zn–1.4Fe–1Sn–0.1Mn), really a high-zinc brass, has high strength combined with excellent wear resistance and is used for clutch disks, extruded shapes, forgings, pump rods, shafting rod, valve stems and welding rod.

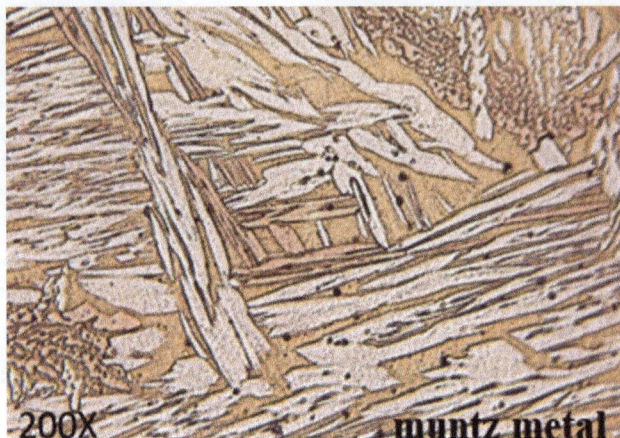

Figure 9.4. The microstructure of Muntz metal (60Cu–40Zn) metal. Copyright George Langford. Reproduced with permission www.georgesbasement.com/.

The above discussion is concerned primarily with wrought brasses, which are mainly binary alloys of copper and zinc. The cast brasses are similar in name to the wrought brasses but usually contain appreciable amounts of other alloying elements. Tin may be present from 1% to 6% and lead from 1% to 10%; some alloys may contain iron, manganese, nickel and aluminum.

9.1.3 Bronzes

The term bronze was originally applied to the copper–tin alloys; however, the term is now used for any copper alloy, with the exception of copper–zinc alloys, that contains up to approximately 12% of the principle alloying element. Bronze, as a name, conveys the idea of a higher-class alloy than brass, and it has been incorrectly applied to some alloys that are really special brasses. Commercial bronzes are primarily alloys of copper and tin, aluminum, silicon or beryllium. In addition, they may contain phosphorus, lead, zinc or nickel.

Tin bronzes
These are generally referred to as phosphor bronzes since phosphorous is always present as a deoxidizer in casting. The usual range of phosphorus content is between 0.01% and 0.5% and that of tin is between 1% and 11%. The copper-rich portion of the copper–tin alloy system is shown in figure 9.5. The β phase forms as the result of a peritectic reaction at 798 °C. At 586 °C, the β phase undergoes a eutectoid reaction to form the eutectoid mixture ($\alpha + \gamma$). At 520 °C, gamma (γ) also undergoes a eutectoid transformation to ($\alpha + \delta$). The diagram also indicates the decomposition of the δ phase. This takes place by a eutectoid reaction at 350 °C forming ($\alpha + \varepsilon$). This reaction is so sluggish that in commercial alloys, the epsilon (ε) phase is nonexistent. The slope of the solvus line below 520 °C shows a considerable decrease in the solubility of tin in the α phase. The precipitation of the δ and ε phases due to this change in solubility is so slow that, for practical purposes, the solvus line is indicated

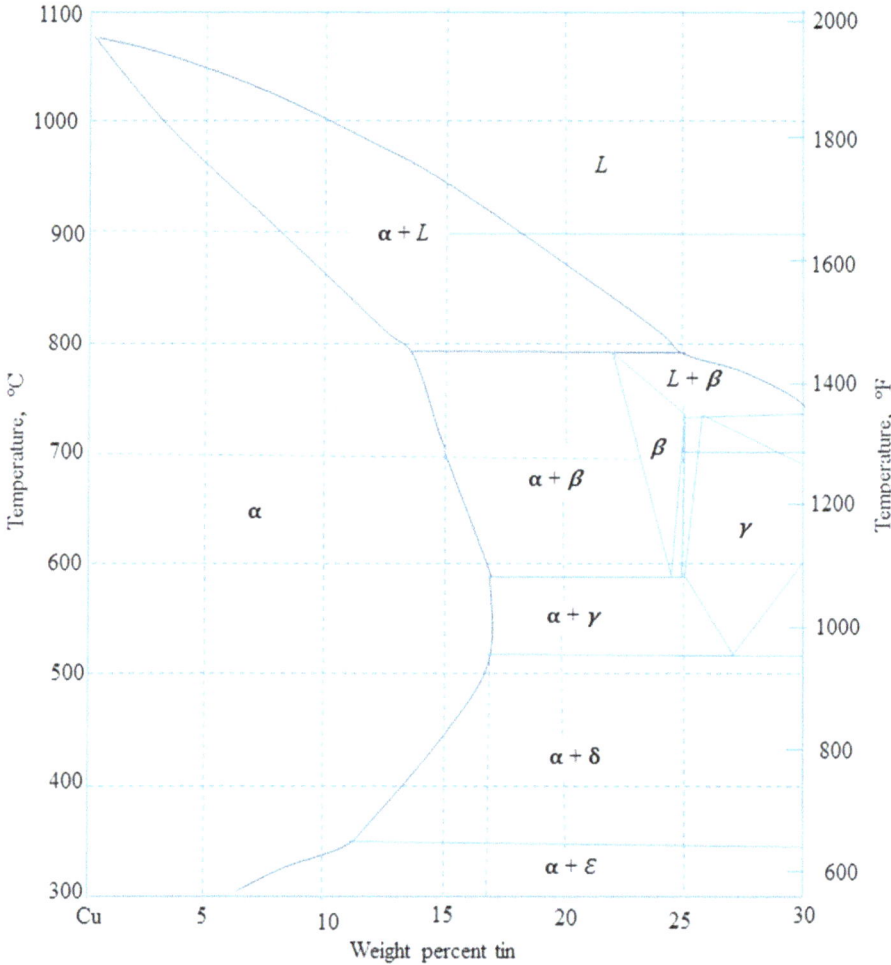

Figure 9.5. Cu-rich portion of Cu–Sn. Reproduced from American Society for Metals (1948) with permission.

by the vertical dotted line below 520 °C. For this reason, slow-cooled cast tin bronzes containing below 7% tin generally show only a single phase, the α solid solution. There is some δ phase in most castings containing over 7% tin. The phosphor bronzes are characterized by high strength, toughness, high corrosion resistance and freedom from season cracking.

Silicon bronzes
The copper-rich portion of the copper–silicon alloy system is shown in figure 9.6. The solubility of silicon in the α phase is 5.3% at 850 °C and decreases with temperature. The eutectoid reaction at 555 °C is very sluggish, so that commercial silicon bronzes, which generally contain less than 5% Si, are single-phase alloys. Silicon bronzes are the strongest of the work-hardenable copper alloys. They have mechanical properties comparable to those of mild steel and corrosion resistance

9-5

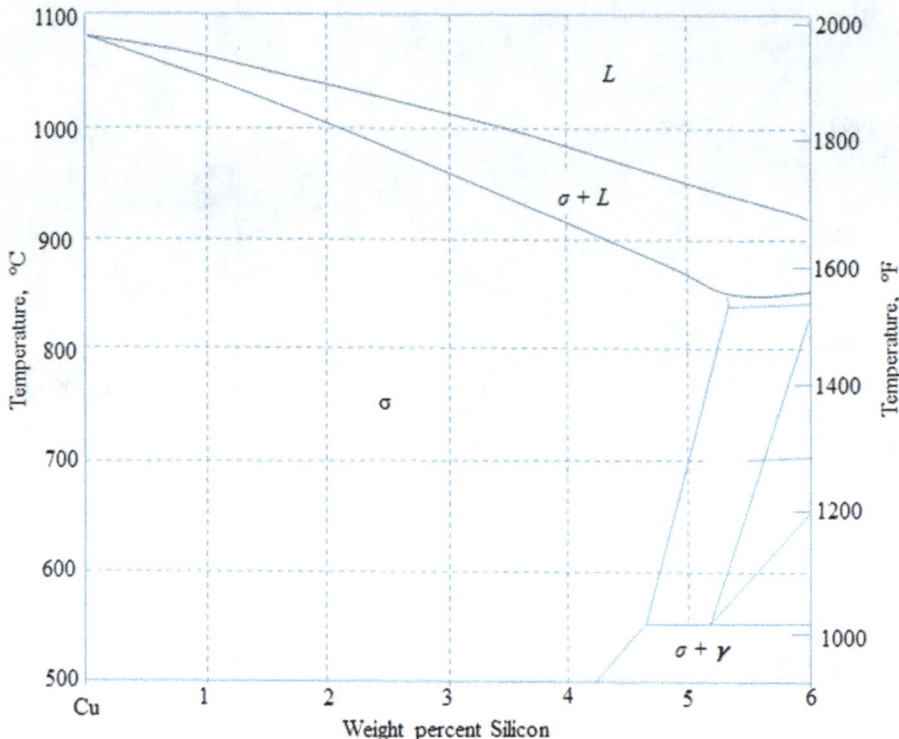

Figure 9.6. The Cu-rich portion Cu–Si phase diagram. Reproduced from American Society for Metals (1948) with permission.

comparable to that of copper. They are used for tanks, pressure vessels, marine construction and hydraulic pressure lines.

Aluminum bronzes
The maximum solubility of aluminum in the α solid solution is approximately 9.5% at 565 °C to form the $(\alpha + \gamma)$ mixture (figure 9.7). Most commercial aluminum bronzes contain 4% and 11% aluminum. Those alloys containing up to 7.5% Al are generally single-phase alloys, while those containing between 7.5% and 11% aluminum are two-phase alloys. Other elements such as iron, nickel, manganese and silicon are frequently added to aluminum bronzes. The single-phase aluminum bronzes show good cold-working properties, with good corrosion resistance to atmospheric and water attack. The $\alpha + \beta$ aluminum bronzes are interesting because they can be heat treated to obtain structures similar to those in steel.

Cupronickels
These are copper–nickel alloys that contain up to 30% nickel. The copper–nickel binary phase diagram shows complete solubility, so all cupronickels are single-phase alloys. They are not susceptible to heat treatment and their properties can be altered only by cold working. The cupronickel alloys have high resistance to corrosion

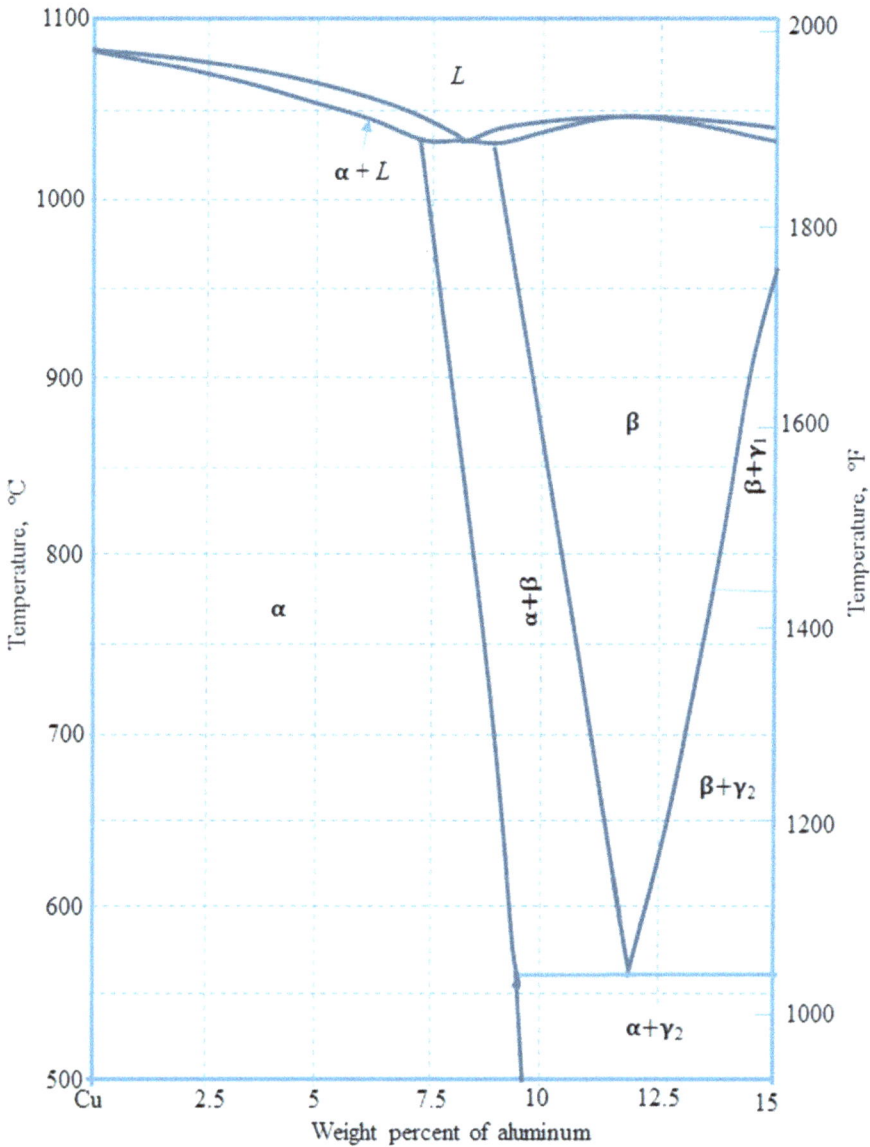

Figure 9.7. Cu–Al phase diagram. Reproduced from American Society for Metals (1948) with permission.

fatigue and also high resistance to the corrosive and erosive action of rapidly moving sea water. They are widely used for condenser, distiller, evaporator and heat exchanger tubes for naval vessels and coastal power plants.

9.2 Aluminum and its alloys

Aluminum is a light metal ($\rho = 2.7$ g c.c.$^{-1}$). It is easily machinable, has a wide variety of surface finishes, good electrical and thermal conductivity, and is nonmagnetic and highly reflective to heat and light. It is a versatile metal; it can be cast, rolled, stamped, drawn, spun, roll-formed, hammered, extruded and forged into many shapes. Aluminum can be riveted, welded, brazed or resin bonded. It is also corrosion resistant. No protective coating is needed, however, it is often anodized to improve the appearance of the surface finish. Al and its alloys have a high strength–weight ratio (high specific strength) owing to their low density. Such materials are widely used in aerospace and automotive applications where weight savings are needed for better fuel efficiency and performance. Al–Li alloys are lightest among all Al alloys and have many wide applications in the aerospace industry.

As described above, aluminum and its alloys possess many attractive characteristics, but the most important characteristic of aluminum is probably its low density, which is about one-third that of steels and copper alloys. Because of this, certain aluminum alloys have a better strength–weight ratio than high-strength steels. Among the many alloying elements added to aluminum, the most widely used are copper, silicon, magnesium, zinc and manganese. These are used in various combinations and in many cases they are used together with other additions to produce classes of age-hardening, casting and work-hardening alloys. All age-hardening alloys contain alloying elements that dissolve in aluminum at elevated (solution treatment) temperatures and precipitate at lower (aging) temperatures. An example of an age-hardening alloy is Al–Cu. Most casting alloys contain silicon, which improves the fluidity and mold-filling capacity of aluminum alloys and reduces their susceptibility to hot cracking and the formation of shrinkage cavities during solidification. Work-hardening alloys frequently contain Mn and Mg, which form a fine dispersion of intermetallic phases and/or impart solid-solution strengthening.

Aluminum alloys are classified into two categories: cast and wrought alloys. Wrought alloys can be either heat-treatable or non-heat-treatable. Alloys are designated by a four digit number. In wrought alloys the first digit indicates the major alloying element. In cast alloys the last digit after the decimal indicates a product from casting (0) or ingot (1) (table 9.1).

9.2.1 Temper designations

The temper designations are as follows:
- F: as fabricated, products with no thermal treatments or strain hardening.
- H: strain-hardened (wrought products) with or without additional thermal treatment.
- H1: strain-hardened only, without thermal treatment.
- O: annealed, recrystallized.
- T: thermally treated with or without strain hardening to produce stable tempers other than F, O or H.
- T3: solution heat-treated and then cold worked.

9.2.2 Heat treatment

The most important heat treating process for non-ferrous alloys is age hardening, or precipitation hardening (figure 9.8). In order to apply this heat treatment, the

Table 9.1. The series of Al in wrought and cast conditions.

Wrought	
Alloy series	Principal alloying elements
1xxx	Minimum 99.00% aluminum
2xxx	Copper
3xxx	Manganese
4xxx	Silicon
5xxx	Magnesium
6xxx	Magnesium and silicon
7xxx	Zinc
8xxx	Other elements
As cast	
1xx.x	Aluminum, 99.00% or greater
2xx.x	Copper
3xx.x	Silicon with copper and/or magnesium
4xx.x	Silicon
5xx.x	Magnesium
6xx.x	Unused series
7xx.x	Zinc
8xx.x	Tin
9xx.x	Other elements

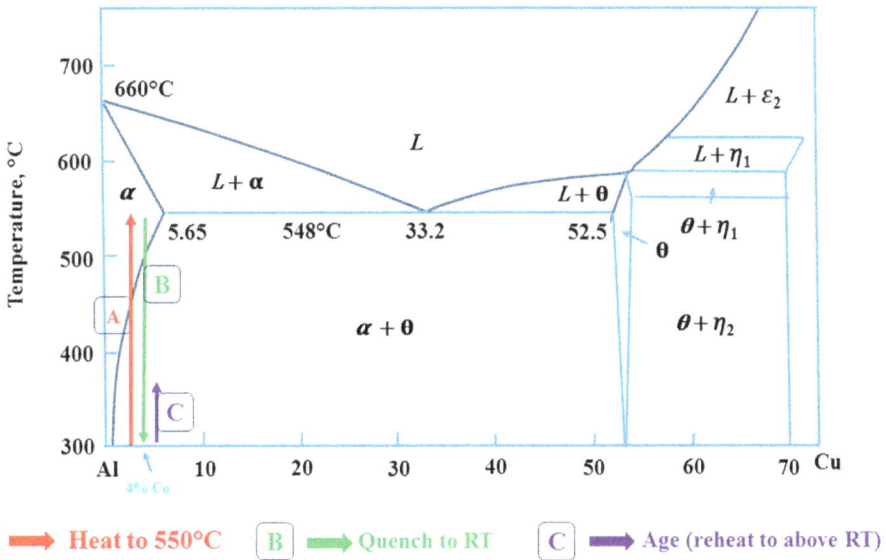

Figure 9.8. The process of precipitation hardening of aluminum.

equilibrium diagram must show partial solid solubility, and the slope of the solvus line must be such that there is greater solubility at a higher temperature than at a lower temperature. The purpose of precipitation-hardening treatment is to improve the strength of the materials. This can be explained with respect to dislocations. The presence of dislocations weakens the crystal leading to easy plastic deformation. Putting hindrance to dislocation motion increases the strength of the crystal. Fine precipitates dispersed in the matrix provide such an impediment. Two stages are generally required in heat treatment to produce age hardening: solution treatment (to dissolve all the alloying elements in the parent matrix) and aging (at subsequent times and temperatures the precipitate can come out and hinder the dislocation motion).

Figure 9.8 explains the precipitation hardening treatment for Duralumin (4% Cu). In this heat treatment process, the main objective is to obtain a fine distribution of precipitates using the cycles A, B and C.

Cycle A: Heat the solid solution (4% Cu alloy) at high temperature to get a single phase region (α).
Cycle B: In this cycle, by quenching high temperature single phase region (α) to room temperature to get a supersaturated solution with increased vacancy concentration.
Cycle C: Ageing can be carried out at room temperature or slightly higher temperature to nucleate a fine precipitate of second phase, to give a strengthening effect for this alloy.

See table 9.2 for the composition, application and mechanical properties of some common aluminum alloys.

9.3 Titanium

Pure titanium melts at 1670 °C and has a low density of 4.51 g c.c.$^{-1}$ (40% lighter than steel and 60% heavier than aluminum). Titanium has high affinity to oxygen, i.e. it is a strong deoxidizer, so can catch fire and cause severe damage. Ti is stronger than Al; its high strength and low weight makes titanium very useful as a structural metal. It has excellent corrosion resistance due to the presence of a protective thin oxide surface film. It can be used as a biomaterial and can be used in elevated temperature components. The main limitation of pure Ti is its lower strength and alloying is often performed to improve this. Oxygen, nitrogen and hydrogen can cause titanium to become more brittle, therefore care should be taken during processing. Titanium can also be cast using a vacuum furnace. Because of its high strength–weight ratio and excellent corrosion resistance, titanium is used in a variety of applications. They are as follows:
- Aircraft body structures and engine parts.
- Sporting equipment.
- Chemical processing.
- Desalination.
- Turbine engine parts.
- Valve and pump parts.

Table 9.2. The composition, applications and mechanical properties of some common aluminum alloys (YS—yield strength, UTS—ultimate tensile strength, %El—% elongation).

Al Ass. No	Composition (wt. %)	Condition	YS (MPa)	UTS (MPa)	% El	Applications
1100	0.12 Cu	Annealed (O)	35	90	45	Food/chemical handling equipment, heat exchangers, light reflectors
3003	0.12 Cu 1.2 Mn 0.1 Zn	Annealed	40	110	30	Utensils, pressure vessels and piping
5052	2.5 Mg 0.25 Cr	Strain-hardened (H32)	195	230	14	Bellows, clutch disks, diaphragms, fuse clips, springs
2024	4.4 Cu 1.5 Mg 0.6 Mn	Heat treated (T4)	325	470	20	Aircraft structures, rivets, truck wheels, screws
6061	1 Mg 0.6 Si 0.3 Cu 0.2 Cr	T4	145	240	22	Trucks, canoes, railroad cars, furniture, pipelines
7075	5.6 Zn 2.5 Mg 1.6 Cu 0.23 Cr	Peak-aged (T6)	505	570	11	Aircraft structures and other highly loaded applications
359.0	7 Si 0.3 Mg	T6	164	228	4	Aircraft pump parts, automotive transmission cases, cylinder blocks
8090	2.0 Li 1.3 Cu 0.95 Mg 0.12 Zr	Heat treated cold worked (T651)	360	465	–	Damage tolerant aircraft structures

- Marine hardware.
- Medical implants and prosthetic devices.
- In bicycles and cars (these are more recent, and increasingly common, applications).

9.3.1 Titanium alloys

Pure Ti exhibits two phases: a hexagonal α-phase at room temperature and a bcc β-phase above 882 °C. The strength of titanium is improved by alloying. The alloying elements are either α or β stabilizers. Elements with an electron–atom ratio <4 are α stabilizers (Al, O, Ga), those with a ratio equal to 4 are neutral (Sn, Zr) and those with a ratio >4 are β stabilizers (V, Mo, Ta, W). ($\alpha + \beta$) two-phase alloys can be obtained with the right proportions of alloying elements. α alloys have low density, moderate strength, reasonable ductility and good creep resistance. Metastable β alloys are heavier, stronger and less ductile than α alloys. Creep strength decreases with increasing β content. ($\alpha + \beta$) alloys show a good strength–ductility combination. The types, compositions, mechanical properties, treatments and applications of some common titanium alloys are given in table 9.3.

9.4 Nickel alloys

Nickel is a high-density, high-strength metal with good ductility and excellent corrosion resistance and high-temperature properties. Ni has many unique properties, including excellent catalytic ability. Nickel is used as a catalyst for fuel cells; nickel–cobalt is seen as a low-cost substitute for platinum catalysts. Two-thirds of all nickel produced goes into stainless steel production. Nickle is also used extensively in electroplating various components in a variety of applications. Ni-based super-alloys are a unique class of materials with exceptionally good high-temperature strength, and creep and oxidation resistance. They are used in many high-temperature applications such as turbine engines. Shape memory alloys—Ni-based (Ni–Ti) and Ni-containing (Cu–Al–Ni) alloys which can return to their original form—are an important class of engineering materials which have found widespread use in many applications. Nickel-containing materials are used in buildings and infrastructure, chemical production, communications, energy (batteries: Ni–Cd, Ni-metal hydrides), environmental protection, food preparation, water treatment and transportation.

9.5 Magnesium

Magnesium is the lightest of the commonly used metals ($\rho = 1.7$ g cm^{-3}). Its melting point is 650 °C and it has an hcp structure. Magnesium is very reactive and readily combustible in air, and can be used as an igniter or fire-starter. The thermal conductivity of Mg is less than that of Al, while their coefficients of thermal expansion are almost the same. Pure Mg has adequate atmospheric resistance and moderate strength. The properties of Mg can be improved substantially by alloying. Magnesium has a favorable atomic size and thus can be alloyed with many elements. The most widely used alloying elements are Al, Zn, Mn and Zr, and Mg alloys come

Table 9.3. The compositions, properties and applications of some Ti alloys.

Alloy type	Comp. wt.% (UNS No)	Condition	YS (MPa)	UTS (MPa)	% El	Applications
Cp Ti	99.1 Ti (R50500)	Annealed	414	484	45	Airframe skins, marine and chemical processing equipment
α	Ti–5Al–2.5Sn (R54520)	Annealed	784	826	16	Gas turbine engine casings and rings, chemical processing equipment
near-α	Ti–8Al–1Mo–1V (R54810)	Annealed	890	950	15	Forged jet engine components: compressor discs, plates, hubs
α–β	Ti–6Al–4V (R56400)	Annealed	877	947	14	Prosthetic implants, airframe components
α–β	Ti–6Al–6V–2Sn (R56620)	Annealed	985	1050	14	Rocket engine cases, airframe structures
β	Ti–10V–2Fe–3Al	Heat-treated (aging)	1150	1223	10	High-strength airframe components, parts requiring uniform tensile stresses

Table 9.4. The compositions, properties and applications of some Mg alloys.

ASTM No	Composition	Condition	YS (MPa)	UTS (MPa)	% El	Applications
AZ31B	3.0 Al 1.0 Zn 0.2 Mn	Extruded	200	262	15	Structure and tubing, cathodic protection
HK31A	3.0 Th 0.6 Zr	Strain-hardened annealed	200	255	9	High-temperature applications (high strength to 315 °C)
ZK60A	5.5 Zn 0.45 Zr	Aged	285	350	11	Forging of maximum strength for aircraft
AZ91D	9.0 Al 0.15 Mn 0.7 Zn	As cast	150	230	3	Die-cast parts for automatic, luggage and electronic devices
AM60A	9.0 Al 0.13 Mn	As cast	130	220	6	Automotive wheels
AS41A	4.3 Al 1 Si 0.35 Mn	As cast	140	210	6	Die-cast parts requiring good creep strength

in both the cast and wrought types. Wrought alloys are available as rods, bars, sheets, plates, forgings and extrusions.

9.5.1 Magnesium alloys

Magnesium alloys are impact and dent resistant, and have good damping capacity; they are thus effective for high-speed applications. Due to its light weight, superior machinability and ease of casting, Mg and its alloys are used in many applications, such as auto-parts, sporting goods, power tools, aerospace equipment, fixtures, electronic gadgets and material handling equipment. Automotive applications include gearboxes, valve covers, alloy wheels, clutch housings and brake pedal brackets. Table 9.4 provides details for some alloys of magnesium.

Further reading

American Society for Metal 1948 *Metals Handbook* (Metals Park, OH: American Society for Metals)

Ashby M F 1998 *Engineering Materials 2: Microstructure, Processing and Design* (Woburn: Butterworth-Heinemann)

Ashby M F 2002 *Engineering Materials 1: An Introduction to Properties, Applications and Design* (Woburn: Butterworth-Heinemann)

Ashby M F 2005 *Material Selection in Mechanical Design* (Woburn: Butterworth-Heinemann)

Askeland D R and Phulé P 2006 *The Science and Engineering of Materials* (Boston, MA: Cengage Learning)

Avner S H 1997 *Introduction to Physical Metallurgy* (New York: McGraw-Hill)

Callister W D 2007 *Callister's Materials Science and Engineering* (Indian Adaptation adapted by R Balasubramaniam) (New Delhi: Wiley)

Raghavan V 2004 *Materials Science and Engineering* 5th edn (Englewood Cliffs, NJ: Prentice-Hall)

Subramaniam A and Balani K (IITK) *Materials Science and Engineering* (e-book) MHRD, India

www.ingramcontent.com/pod-product-compliance
Lightning Source LLC
Chambersburg PA
CBHW081534220326
41598CB00036B/6436